Francis Bacon and the Heroic Archetype

Francis Bacon and the Heroic Archetype: The Greatest Story Never Told

Ryan Murtha

RESOURCE *Publications* • Eugene, Oregon

Francis Bacon and the Heroic Archetype
The Greatest Story Never Told

Wipf & Stock
An Imprint of Wipf and Stock Publishers
199 W. 8th Ave., Suite 3
Eugene, OR 97401

www.wipfandstock.com

PAPERBACK ISBN: 979-8-3852-7877-0
HARDCOVER ISBN: 979-8-3852-7878-7
EBOOK ISBN: 979-8-3852-7879-4

Contents

Part I: Preliminary

Part II: The Greatest Story Never Told

Part III: Essays

Francis Bacon and the Hero Pattern (after Raglan)

Raglan Criterion	Hero Pattern Description	Francis Bacon (Structural Correspondence)
1	The hero is born to a royal virgin	Bacon is born into the immediate orbit of royal power, amid widespread rumors concerning Elizabeth I.
4	The circumstances of his conception are unusual	Bacon's birth coincides with diplomatic anxiety and rumors of illicit intimacy between Elizabeth and Robert Dudley. The unresolved death of Amy Robsart. Conditions of secrecy and danger, rather than domestic stability.
8	Reared by foster-parents	Bacon is raised within powerful households (Bacon, Cecil, the court) and educated not as a private child, but as a political instrument within a wider system of patronage and service.
10	On reaching manhood, returns to his future kingdom	After spending three years in France, Bacon returns to England at age eighteen, following his father's death.
11	Victory over a king, giant, or beast	Bacon's overthrow of Aristotelian scholasticism in the *Novum Organum* functions as a symbolic defeat of the reigning intellectual sovereign—the "king" of the schools.
13	Becomes king	Temporary regent of England during James I's visit to Scotland in 1617; quasi-regal authority as Lord Chancellor. More enduringly, he becomes "king" in the domain of knowledge by legislating method rather than ruling territory.

Raglan Criterion	Hero Pattern Description	Francis Bacon (Structural Correspondence)
15	Prescribes laws	As Solicitor General, Attorney General, and Lord Chancellor; philosophically, he prescribes laws for inquiry itself, articulating a moral constitution for science.
17	Driven from the throne and city	Bacon's fall from office in 1621 results in his formal banishment from London and exclusion from political life, an archetypal expulsion following the giving of law.
18	Meets with a mysterious death	Bacon's death, following his cold-weather experiment, has long been treated as emblematic and possibly apocryphal, constructed to frame Bacon as a martyr to science, like Pliny the Elder.
19	Often dies at the top of a hill	Dies at Highgate; as the name implies, an elevated section of London.
20	His children, if any, do not succeed him	Bacon leaves no legitimate heirs; his intellectual legacy passes institutionally rather than genealogically, consistent with the heroic founder rather than dynastic king.

History of the Royal Society (1667) with Bacon as *Artium Instaurator,*
"Restorer of the Arts"

Introduction

When Francis Bacon revised his will in December 1625, he was a disgraced former lord chancellor, heavily in debt, politically isolated, and keenly aware that his reputation in England was unlikely to recover soon. "For my name and memory," he wrote, "I leave it to men's charitable speeches, and to foreign nations, and the next ages." This was not an exercise in vanity or a plea for absolution; Bacon understood that the intellectual project to which he had devoted his life, the reformation of knowledge itself, could not be judged fairly by his contemporaries. He therefore addressed himself, deliberately and without sentimentality, to posterity.

Yet there is more to the story. The earliest biographical account of Bacon was published in 1657 by William Rawley, who had served as Bacon's chaplain, amanuensis, and literary executor. It begins "Francis Bacon, the glory of his age and nation, the adorner and ornament of learning, was born at York House, or York Place, in the Strand, on the two-and-twentieth day of January, the year of our Lord 1560 [1561 N.S.]." This is a deliberate anachronism, and much more significant than may be supposed. York *House* was the London residence of Sir Nicholas and Lady Anne Bacon; after Sir Nicholas acquired it in the mid-16th century, it became firmly associated with the Bacon family. York *Place* was originally the palace of the Archbishops of York. Cardinal Wolsey occupied it until his fall in 1529; whereupon Henry VIII seized York Place, renamed it Whitehall, and massively expanded it. So by Bacon's birth, York Place as such no longer existed, it had already become Whitehall Palace, the principal royal residence.

This book is both speculative theology and revisionist history. Its core thesis is that the heroic archetype manifested in the person of Francis Bacon. The interpretive framework draws from a body of literature on the archetype—a word Bacon imported into English, incidentally[1]—principally Lord Raglan, but also Otto Rank, Erich Neumann, Joseph Campbell, and

[1] *OED*, s.v. "archetype, n." (earliest evidence 1605, Bacon); *Advancement of Learning* (1605), I.VI.1. "First therefore let us seek the dignity of knowledge in the archetype or first platform, which is in the attributes and acts of God"

others. Although I make use of Raglan's list of heroic attributes, I do not accept his ritualist explanation of their origin or meaning; the list is employed here strictly as a descriptive instrument, a means of identifying recurring narrative and biographical patterns, without endorsing the theory that such patterns necessarily derive from ritual practice. Instead, I tend to align with interpretative currents that see these patterns as manifestations of a deeper, divine logic woven into the fabric of history. In this sense, Bacon's life can be read as part of a larger tapestry of mythic meaning, an intersection of the human and the divine.

Bacon is an unusual instance because he is, in many respects, the least mythical subject imaginable. Few public lives in early modern England are so copiously documented, so densely embedded in institutions, so exposed to the dry daylight of letters, offices, commissions, speeches, and lawsuits. And yet, the more Bacon becomes legible to history, the more his career begins to resemble something older than history: the lawgiver who provokes rivalry, suffers expulsion, and is finally made to bear a communal burden larger than any private fault. Myth does not replace biography; but at certain pressure points, biography begins to behave like myth.

I proceed from the conviction that Bacon cannot be understood through conventional biography alone. His life unfolded in a world where truth was elusive and could not be spoken plainly without hazard. To read Bacon properly, one must learn to read as his age required: attentively, obliquely, and with an eye for what is not said. In *The Wisdom of the Ancients,* his book of allegorical interpretations of Greek mythology, Bacon begins with a preface outlining his reasons for seeking deeper meaning in the stories: "My judgment is, that a concealed instruction and allegory was originally intended in many of the ancient fables":

> The argument of most weight with me is this, that many of these fables by no means appear to have been invented by the persons who relate and divulge them, whether Homer, Hesiod, or others; for if I were assured they first flowed from those later times and authors that transmit them to us, I should never expect anything singularly great or noble from such an origin.

He pointedly tells us that he does not believe the works of Homer and Hesiod were composed by their purported authors; is this a sly allusion to Shakespeare?

To speak of myth here is not to imply fabrication. It is to propose that patterns of disclosure, spiritual, political, and psychological, may emerge in lives as well as in stories. Revelation is rarely a bolt from the sky; it is more often mediated and partially veiled, because what is disclosed is too complex to be communicated simply, and too consequential to be endured nakedly. Myth, adjacent to art, is a form that lets truth approach us without overwhelming us.

This is particularly fitting for the English court of the late Tudor and early Stuart world, a theatre in which power was not only exercised but staged, where sovereignty depended on the management of appearances, and where the cost of speaking too directly could be professional ruin or worse. Bacon himself treats inherited myth not as childish fantasy, but as encoded philosophy, a symbolic treasury in which earlier ages stored what they knew, including what could not prudently be said outright. His method there, the patient extraction of meaning from fable, licenses the method here: to read patterns in public life not as mere coincidences, but as possible signatures of intelligible order.

At the center of that order stands Elizabeth, England's Virgin Queen; and yet, in the deeper logic of political imagination, something more than virgin. The cult of virginity was not merely a personal detail or a moral badge; it was an instrument of statecraft, a theology of sovereignty. Elizabeth is repeatedly staged as inviolate and self-sufficient, the spouse of the realm, the image of a nation that requires no consort and no rival source of legitimacy. But virginity in political form tends to generate its own shadows. When a queen must be immaculate, the court becomes a machine for suppressing inconvenient intimacy and transmuting private realities into public myth.

Within this charged atmosphere Bacon's story takes shape. He presents himself, memorably and audaciously, as one who has taken all knowledge to be his province. He does not seek a kingdom of land; his library is dukedom enough, he claims a kingdom of mind. His dominion is wisdom, his law is epistemic law. If the heroic archetype has a modern form suited to an age of letters and institutions, Bacon is a strong candidate: a contemplative hero whose feat is neither conquest nor slaughter, but a re-founding of method, a re-chartering of how the human mind may lawfully approach nature. "For," he wrote, "as the works of wisdom surpass in dignity and power the works of strength, so the labours of Orpheus surpass the labours of Hercules." Here the Christian resonance becomes difficult to miss: not a king of this world, yet a kind of sovereignty nonetheless.

The political drama around him sharpens the pattern. The Essex rebellion of 1601 was an unsuccessful coup attempt on Elizabeth's throne led by Robert Devereaux, the Earl of Essex. Bacon's role in prosecuting a friend and former benefactor has often been narrated as personal treachery or careerism; however, another reading is possible, one that does not excuse, but re-contextualizes. Rebellions against the sovereign frequently assume a theological posture: not merely opposition, but usurpation, a bid to seize legitimacy by force. The Essex episode can be read, archetypally, as a Luciferian rehearsal within the state: the favorite who overreaches, the bright one who cannot remain subordinate, the attempted seizure of the throne's aura. In such a drama, someone must become the instrument of judgment. Bacon's part, taken reluctantly, casts him as the agent through whom the system purges its insurgent angel.

And yet the purge does not end with Essex. The deeper logic of sacrifice often turns, in time, upon the lawgiver himself. Bacon's fall, swift, humiliating, formally codified, bears the unmistakable contour of scapegoating: a single figure made to carry a cluster of institutional sins and factional tensions, so that the prestige of the whole may be preserved. It is here that Raglan's pattern becomes most unsettling, not because it proves ritual origins, but because it exposes the grammar of political redemption: the community restores itself by expelling the one who becomes too visible a point of convergence for its contradictions.

Some readers will wish to press further, into questions of origin. Was Bacon, in some sense, a hidden prince? Are there buried genealogies, secret births, suppressed paternities that would align the outer facts with the inner pattern? This book will treat such speculations as what they are: hypotheses, not certainties. Here, their value is not chiefly forensic, but interpretive. It belongs, as a possibility, to the atmosphere of the tale; it need not be asserted as dogma to do explanatory work. In any case, Francis Bacon was raised within the immediate orbit of the court, at the very center of the state.

The governing claim of this introduction, then, is simple: Bacon's life is best approached at two levels at once. On one level, it is a career in offices, patronage networks, legal reform, parliamentary maneuver, and courtly peril; an English story with hard dates and hard documents. On another, it appears as a drama through which something larger than biography becomes legible: the way sovereignty manufactures purity; the way rebellion assumes theological form; the way lawgiving invites scapegoating; the way the mind's kingdom provokes the world's resentment. If myth is, as many have argued, a mode of divine self-disclosure, symbolic truth given in story, then it may disclose itself not only in ancient fables, but in the lives of those who, by temperament or by destiny, become mirrors for the deep logic of history. What follows is offered in that spirit: neither credulous nor reductive; neither scandal-mongering nor disinfected; but an attempt to read Bacon with both eyes open—one on the archive, and one on the pattern.

Bacon as scientific Moses

Francis Bacon's political rise and fall follow with remarkable fidelity the pattern identified by Raglan in heroic biography: the figure who approaches sovereignty, prescribes law, and is then expelled—remembered less for rule than for legislation, less for power than for order. Bacon never wore a crown, yet he came close to symbolic kingship: as Lord Chancellor he stood second only to James I, and during the king's absence in 1617 he briefly functioned as effective regent. On that occasion he appeared in full ceremonial purple, a color restricted by law to royalty (a detail not lost on contemporaries).

Yet Raglan's hero does not rule as history's conquerors rule. Once elevated, his reign is curiously uneventful; he founds no dynasty, builds no empire, and leaves behind no lineage. Instead, his lasting memorial is a body of law, often attributed to him personally, though in reality codified from long historical development. Bacon's career conforms exactly to this paradox. He did not invent English law, but he acted repeatedly as a lawgiver in the deepest sense, seeking to regulate not merely courts and statutes, but the future conduct of knowledge itself.

This role was recognized early and explicitly. In *Minerva Britanna* (1612), Bacon is compared to Solon, the archetypal Greek lawgiver. More significant still is the language used by the founders of the Royal Society. In Thomas Sprat's *History of the Royal Society* (1667), Bacon is cast explicitly in Mosaic terms:

> Bacon, like Moses, led us forth at last.
> The barren wilderness he past,
> Did on the very border stand
> Of the blest promis'd land

> And from the mountain's top of his exalted wit
> Saw it himself and shew'd us it.

The analogy is apt. Moses leads his people out of bondage, receives the law, and dies before entering the promised land. Bacon conceived his role in precisely these terms; he framed the method, delineated the boundaries, and oriented science toward a future order he himself would never inhabit. Crucially, Bacon did not understand this task as morally neutral; he insists that knowledge, if unbounded, is dangerous. He likens the emerging science to a powerful and unpredictable flood, requiring a "strong and sound bank" to guide its course. That bank is ethical constraint; knowledge, he insists, must be used for "the benefit and relief of the state and society of man." His greatest hope was that his efforts would, in time, "overcome the immeasurable helplessness and poverty of the human race," and "make you peaceful, happy, prosperous, and secure." Bacon does not merely open the fountains of power, he attempts to regulate their flow.

Raglan's hero, having prescribed laws, is expelled from the city. Bacon's impeachment, engineered by his long-standing rival Sir Edward Coke, ended his public career with stunning abruptness. Formally convicted on twenty-three counts of corruption, he was fined, imprisoned, barred from office, excluded from Parliament, and banished from the court's physical presence. Although most of these penalties were quietly annulled within days by James, the symbolic outcome was irreversible: Bacon was driven from the center of power.

The historical record strongly suggests that this was less a reckoning with venality than a political sacrifice. Despite his acknowledged talents as an orator, Bacon's law practice earned far less than his contemporaries; he lived perpetually in debt and showed little evidence of financial rapacity. More tellingly, as Lord Chancellor he had begun dismantling the fee-based judicial economy that enriched common-law judges, advocating instead for salaried offices paid by the state. By clearing massive case backlogs and making Chancery faster and more accessible, he directly threatened entrenched interests, particularly Coke's. In heroic terms, the lawgiver who reforms the system must be destroyed by those whose power depends on the old order.

Bacon's death completes the pattern. He dies under ambiguous circumstances, framed almost immediately as a martyr to science, and he dies, fittingly, on elevated ground at Highgate overlooking London. He leaves no children. His legacy passes not through blood, but through institutions, manuscripts, and future generations. Like Moses, he is permitted to see the promised land but not to enter it. The science he envisioned would flourish only after his death; and, crucially, after the erosion of the moral boundaries he tried to impose.

Aristotle and the overthrow of the father

Across twentieth-century myth theory, a persistent structure emerges: the hero comes into being through the symbolic displacement of a paternal authority whose rule, once life-giving, has become rigid and sterile. This "overthrow of the father" is rarely literal and almost never moralized. The father is not evil; he is exhausted by time; his authority rests on inheritance, rather than responsiveness to a changing world. The heroic act is therefore lawful succession, a transfer of sovereignty from past to future, from repetition to renewal. When this pattern is translated from family or political myth into intellectual history, the "father" need not be a person at all. He may be a canon, a method, or an epistemic regime that has hardened into unquestioned rule.

By the late sixteenth century, Aristotle occupied precisely this position. No longer one thinker among many, he had become the sovereign of European learning, governing philosophy through inherited categories, commentary, and deduction. It is against this backdrop that Francis Bacon's intervention assumes its mythic force. The *Novum Organum* does not merely criticize Aristotle; it supplants him. Bacon performs a symbolic regicide; authority shifts from paternal tradition to disciplined experience, from top-down deductive logic to bottom-up inquiry. In heroic terms, the old king is not destroyed but succeeded; knowledge is re-grounded without the cosmos collapsing. For an age struggling to move from inheritance to discovery, Bacon's greatness lies not only in the content of his philosophy, but in the heroic effort with which he removed a father and left the world intelligible. He even suggested, at least to himself, that he might have been somehow more than human:

> Now if the utility of any single invention so moved men, that they accounted more than man him who could include the whole human race in some solitary benefit, that invention is certainly much more exalted, which by a kind of mastery contains within itself all particular inventions, and delivers the mind from bondage, and opens it a road, that under sure

> and unerring guidance it may penetrate to whatever can be of novelty and further advancement.[2]

Bacon and the foundations of binary computing

In *De Augmentis Scientiarum,* Bacon makes a remark that has often been dismissed as a cryptographic curiosity. In fact, it articulates a far deeper principle: meaning can be conveyed by any medium whatsoever, provided it admits of a stable twofold difference:

> Neither is it a small matter these cypher-characters have, and may perform: for by this art a way is opened, whereby a man may express and signify the intentions of his mind, at any distance of place, by objects which may be presented to the eye, and accommodated to the ear: provided those objects be capable of a twofold difference only.

Bacon's insight is not about letters or secrecy, but about encoding itself. Meaning does not inhere in symbols, sounds, or marks as such; it resides in difference, presence and absence, high and low, light and dark, sound and silence. Once this abstraction is made, content becomes independent of its material carrier, a separation that lies at the heart of digital representation.

Bacon anticipates the central intuition behind modern computing and information theory; his bi-literal cipher is the first binary system of encoding information. Long before binary arithmetic, symbolic logic, or formal machines, he identifies the minimal condition under which information can exist at all: consistent, repeatable distinction. Later figures would formalize and mechanize this insight, but Bacon supplies its philosophical ground. The bi-literal cipher demonstrates the principle that the same message may pass unchanged through typography, formatting, or any other controllable variation. Bacon thus appears, unexpectedly, as a philosopher of media abstraction: one who grasped that thought, once reduced to structured difference, becomes portable across media, embodiment, and time. It is in this sense, not as an engineer but as a metaphysician of information, that Bacon stands at the beginning of the intellectual trajectory leading to modern computing.

---

[2] *Thoughts concerning the Interpretation of Nature,* Tr. Basil Montagu *The Works of Francis Bacon* London: William Pickering 1834

Boon or bane: Bacon's legacy in retrospect

> What I have been able to do is to give it, as I hope, a not contemptible start. The destiny of the human race will supply the issue, and that issue will perhaps be such as men in the present state of their fortunes and their understandings cannot easily grasp or measure. For what is at stake is not merely a mental satisfaction, but the very reality of man's wellbeing, and all his power of action.
>
> —*Novum Organum*

The world we inhabit, for better and for worse, bears the imprint of Bacon's intervention; any attempt to recover him as a heroic, world-renewing figure must confront the ambivalence of his legacy. The triumph of consciousness over inherited ignorance, so central to Bacon's self-understanding and to the heroic archetype itself, did not unfold without cost. The methods he helped inaugurate proved extraordinarily powerful; but power, once unleashed, does not remain tethered to the intentions of its originator. In hindsight, Bacon's vision stands at the threshold of modernity's great promise, as well as its great anxiety: that the conquest of nature through knowledge might liberate humanity, or might instead estrange us from meaning, restraint, and wisdom.

From one perspective, Bacon's legacy is unmistakably a boon. The reorientation of inquiry toward experiment, collaboration, and cumulative progress shattered intellectual stagnation and enabled advances that transformed medicine, technology, and material conditions on a global scale. His insistence that knowledge should be fruitful, rather than merely contemplative, helped break the monopoly of inherited authority and restored a sense of agency to human reason. In this respect, Bacon's project fulfilled the heroic function described earlier: it renewed a world whose symbolic and epistemic forms had become rigid, opaque, and unresponsive to lived reality.

Yet it is equally clear that Bacon's name has come to symbolize, for many, the darker trajectory of modern thought. The accusation is not new; in his annotated copy of Bacon's *Essays*, William Blake wrote "The Prince of darkness is a Gentleman and not a Man, he is a Lord Chancellor." For Blake, Bacon and Newton were of a piece; the same methodological clarity that dissolved superstition also encouraged a view of nature as inert matter to be mastered, quantified, and exploited. Bacon is often held responsible, fairly or not, for a technocratic worldview in which efficiency eclipses wisdom and means overwhelm ends. The hero who slays the monster of darkness can, in this telling, become the unwitting architect of a new shadow: a civilization rich in power yet uncertain of purpose.

What is frequently missed in this retrospective judgment, however, is the degree to which Bacon himself anticipated the danger. His repeated emphasis on moral discipline, intellectual humility, and the purification of intention suggests that he did not envision method as self-justifying. The idols he warned against were not only scholastic errors, but distortions born of ambition, vanity, and collective appetite. Bacon's hope was that a reformed intellect, properly governed, might restore a lost harmony between humanity, nature, and divine order. That later generations severed method from metaphysics, and power from responsibility, cannot simply be laid at his feet.

Seen through the lens of the heroic archetype, this ambivalence is not anomalous, but characteristic. The hero opens a path that others must walk, often without fully understanding its original meaning. Bacon's legacy, then, should not be judged solely by the outcomes of modernity, whether triumphant or catastrophic, but by the nature of the transformation he sought to initiate. His was an attempt to bring consciousness to bear on inherited darkness; not to abolish mystery, but to discipline ignorance. Whether that effort has ultimately been a boon or a bane remains an open question, and perhaps must remain so. What can be said with confidence is that Bacon stands at the origin of a world we still inhabit, and of a struggle between knowledge and wisdom that has yet to be resolved.

A lost voice on power and knowledge

Since the middle of the twentieth century, and especially since Hiroshima, it has become increasingly difficult to maintain the older confidence that scientific power is self-justifying. The problem is not science itself, but the speed with which technical capacity now outruns ethical deliberation. We possess unprecedented means of transformation—of nature, of bodies, of societies, without any comparable consensus about restraint, purpose, or limits. This unease is widespread, crossing political, religious, and cultural boundaries. One need not reject science to feel it; indeed, it is often felt most acutely by those who understand its power best.

What is less often noticed is that this anxiety is not new. It was articulated with remarkable clarity at the very moment when modern science first began to take shape. Francis Bacon, long remembered as a herald of experimental method and technological progress, was also among the earliest thinkers to insist that knowledge, precisely because of its power, must be morally governed. He warned repeatedly that the unchecked pursuit of mastery would deform both the knower and the society that empowered him. For Bacon, science was not merely a tool for domination or utility; it was a vocation that demanded humility, ethical orientation, and self-limitation.

And yet this Bacon is largely absent from modern memory. Over the centuries, his method was adopted, refined, and amplified, while his cautions were quietly set aside. His reputation was simplified, his character narrowed, and his deeper concerns treated as marginal or inconvenient. What survived was a flattened image: Bacon as a forerunner of technique rather than a legislator of conscience. In losing that dimension of his thought, we may have lost more than a historical nuance. We may have lost a way of thinking that could still help us navigate the moral consequences of our own success.

The argument of this book is not that Bacon offers solutions to contemporary dilemmas, nor that he should be revered uncritically. It is that he represents an untapped resource, a voice from the origin of modern science that recognized, earlier than most, the danger of power without restraint. To recover Bacon in this fuller sense is not to retreat from modernity, but to ask whether something essential was discarded along the way.

There is a sadness in this realization, but also a possibility. If Bacon's warnings were ignored once, they need not be ignored forever. At a moment when the ethical governance of technological power has become one of the central questions of human survival, it may be worth listening again to a thinker who believed that the future of science depended not only on what we could do, but on what we chose not to do.

The sadness deepens when one realizes that Bacon is not merely a figure misjudged by history, but a resource we have abandoned, one uniquely suited to our present predicament. He stands at a moment when knowledge, power, and ethics had not yet been fully severed from one another. He saw what was coming with remarkable clarity: the exponential growth of technical capacity, the temptation to equate mastery with wisdom, and the danger of a science unmoored from moral orientation.

What makes Bacon so valuable now is precisely what made him inconvenient later. He did not imagine science as a purely instrumental enterprise. He conceived it as a moral vocation, one that required restraint, humility, and an ongoing reckoning with human fallibility. He anticipated, because he personally witnessed, the crises that arise when power outruns conscience.

In discarding Bacon, modernity did not simply refine or improve upon his project; it narrowed it. His method was retained, because it was useful. His ethical architecture was allowed to erode, because it was limiting. Over time, the figure who insisted that "all knowledge is to be limited by religion" and ordered toward "the relief of man's estate" became an embarrassment: too serious, too demanding, too resistant to triumphalist narratives of progress.

The result is a peculiar historical irony. At precisely the moment when we most need a way of thinking that can hold together scientific power and moral responsibility, Bacon is treated as obsolete, transitional, or ethically compromised. We consult him as a historical precursor, not as a living interlocutor. And in doing so, we overlook the possibility that he offers not answers, but a framework for asking questions about limits, purposes, and the kind of future we are actually building.

This is why Bacon's marginalization feels less like the correction of an error than the quiet loss of an ally. He does not tell us what to do. He tells us how to think when power is growing faster than wisdom. That is not an antiquarian concern. It is, perhaps, the defining concern of our time.

1697 engraving of Elizabeth I by Cornelis Vermeulen

1. Francis Bacon and the hero archetype

Classical mythology locates heroism in physical deeds: slaying monsters, founding cities through blood, and displacing rivals through force. Francis Bacon, standing at the threshold of modernity, relocates heroism from action to understanding. Where the mythic hero liberates the community by killing, Bacon's hero liberates humanity by knowing; a hero not of might, but of the mind. The enemy is no longer a beast, but ignorance; the decisive act is no longer sacrifice, but method. In this sense, Bacon offers a post-mythic form of heroism; knowledge becomes not a prelude to power, but a non-violent substitute for it.

In Joseph Campbell's formulation, the true hero is not merely an adventurer or conqueror, but a world-renewing figure; the hero's task is accomplished not for private glory, but for the revitalization of a culture whose symbolic and intellectual forms have grown rigid, exhausted, or opaque. Campbell gave the hero pattern its most familiar form: he departs, undergoes ordeal, and returns bearing a boon. Bacon appears as a self-conscious bearer of such a boon; his language, when writing about himself, sometimes seems grandiose, even quasi-messianic at times. His ambition was not to deliver a single invention or discovery, but to kindle a reform of knowledge for "the benefit and relief of the state and society of man."

Campbell emphasizes that such figures appear at moments of historical blockage, when inherited structures no longer mediate meaning effectively between humanity and reality. Bacon's project aligns precisely with the mythic function Campbell describes. The Great Instauration is an act of renewal, a clearing away of dead forms so that knowledge may once again serve life. His visionary role consists in recognizing that the world itself must be re-approached, re-read, and re-founded, lest civilization continue to mistake tradition for truth.

Carl Jung defines the central labor of the hero with clarity: "The hero's main feat is to overcome the monster of darkness: it is the long-hoped-for and expected triumph of consciousness over the unconscious."[3] In this sense, Francis Bacon's project can be seen as a paradigmatic heroic undertaking. His lifelong campaign to reform knowledge, to tear it away

---

[3] *The Collected Works of C. G. Jung, Volume 9 (Part 1): Archetypes and the Collective Unconscious*. Princeton University Press, 1969

from inherited authority, scholastic habit, and unexamined tradition, was nothing less than an attempt to bring the unconscious assumptions of Western thought into the light of disciplined inquiry. The *idola* Bacon identified were not merely intellectual errors, but collective shadows: habits of mind so deeply ingrained that they masqueraded as truth itself. To overcome them required a new method, a new discipline of attention, and a willingness to confront the darkness of ignorance without illusion or consolation.

Bacon understood this task not as an abstract philosophical exercise, but as a vocation. Again and again in his writings he presents himself as one called, even compelled to this labor, as though he had been born to initiate a transformation in humanity's relation to nature and to knowledge itself. Here he should be quoted at length:

> Whereas, I believed myself born for the service of mankind, and reckoned the care of the common weal to be among those duties that are of public right, open to all alike, even as the waters and the air, I therefore asked myself what could most advantage mankind, and for the performance of what tasks I seemed to be shaped by nature. But when I searched, I found no work so meritorious as the discovery and development of the arts and inventions that tend to civilize the life of man . . . Above all, if any man could succeed not merely in bringing to light some one particular invention, however useful, but in kindling in nature a luminary which would, at its first rising, shed some light on the present limits and borders of human discoveries, and which afterwards, as it rose still higher, would reveal and bring into clear view every nook and cranny of darkness, it seemed to me that such a discoverer would deserve to be called the true Extender of the Kingdom of Man over the universe, the Champion of human liberty, and the Exterminator of the necessities that now keep men in bondage. Moreover, I found in my own nature a special adaptation for the contemplation of truth. For I had a mind at once versatile enough for that most important object—I mean the recognition of similitudes—and at the same time sufficiently steady and concentrated for the observation of subtle shades of difference. I possessed a passion for research, a power of suspending judgment with patience, of meditating with pleasure, of assenting with caution, of correcting false impressions with readiness, and of arranging my thoughts with scrupulous pains. I had no hankering after novelty, no blind admiration for antiquity. Imposture in every shape I

> utterly detested. For all these reasons I considered that my nature and disposition had, as it were, a kind of kinship and connection with truth.[4]

In the Jungian sense, then, Bacon's Great Instauration is not simply a methodological reform; it is the heroic drama of consciousness asserting itself against the weight of inherited darkness, seeking not domination for its own sake, but the restoration of truth through clarity, patience, and light.

Erich Neumann addressed the monomyth in *The Origins and History of Consciousness,* drawing out a possibility already latent in Freud's *Moses and Monotheism*: the hero myth is not merely about exceptional individuals, but about the self-unfolding of the divine image in history. The hero is not simply a servant of the god-image; he is the vehicle through which a new configuration of the divine becomes historically operative. Neumann writes:

> The hero, as bringer of the new, is the instrument of a new manifestation of the father god. In him the patriarchal gods struggle against the Great Mother, the invaders' gods against the indigenous gods, Jehovah against the gods of the heathen. Basically it is a struggle between two god images or sets of gods, the old father-god defending himself against the new son-god, and old polytheistic system resisting usurpation by the new monotheism, as is exemplified by the archetypal wars of the gods.

The struggle Neumann describes is simultaneously mythological, theological, and historical. The hero mediates a transition between divine regimes: from maternal, cyclical, earth-bound orders to paternal, linear, law-giving ones; from many gods embedded in place to one God asserting universal sovereignty. The hero's trials, exile, conflict with established powers, and eventual fall or sacrifice are not incidental narrative flourishes but reveal the cost of reconfiguring the sacred itself.

Read this way, the hero archetype becomes a structural analogue to monotheism. Just as monotheism concentrates divine authority into a single principle, the hero concentrates mythic action into a single figure. The hero bears the burden of differentiation: separating law from nature, transcendence from immanence, command from fertility. This helps explain why the hero so often suffers rejection, persecution, or erasure. If the hero represents the emergence of a new god-image, resistance to the hero is also resistance to that transformation.

---

[4] Durant, Will *The Story of Philosophy*. New York: Simon & Schuster, 1926. pp. 119-120. Quoting Abbot, *Francis Bacon* (1885).

A striking historical instance of this logic is Akhenaten, the first major monotheistic ruler. His attempt to impose exclusive devotion to the Aten was not merely a religious reform, but a radical act of differentiation: the concentration of cosmic meaning into a single, abstract principle, mediated by a single figure. The backlash was correspondingly severe. After his death, Akhenaten was subjected to an unusually thorough *damnatio memoriae*: his name was chiseled from monuments, his images defaced, his capital abandoned, and the old polytheistic order restored with ritual insistence. This was not the quiet fading of a failed ruler, but an active effort to undo a transformation judged intolerable. Akhenaten's erasure thus clarifies the structural stakes of the hero archetype: when a figure embodies a new god-image, one that reorders authority, knowledge, and mediation, the rejection that follows aims not only at the person, but at the principle he represents. Bacon has suffered a similar fate.

Neumann's formulation also clarifies why heroic narratives persist even in secularized cultures. The hero continues to appear as lawgiver, founder, or reformer, not because societies consciously replay ancient myths, but because the pattern of divine self-manifestation has not been exhausted. The hero story, in this sense, does not merely reflect belief; it carries it forward, translating metaphysical change into historical form.

Neumann stresses that the hero myth is never about private biography, but is rather about transpersonal transformation:

> The hero myth is never concerned with the private history of an individual, but always with some prototypal and transpersonal event of collective significance ... Although they appear as inner events, the victory and transformation of the hero are valid for all mankind; they are held up for our contemplation, to be lived out in our own lives, or at least re-experienced by us.

For Neumann, the hero's struggle enacts a decisive shift in the psychic structure of a culture: the emergence of a new relation between consciousness and the unconscious that has collective consequences. The hero does not merely defeat the monster; he integrates what was previously dark, chaotic, or unarticulated into an expanded order of meaning. Bacon's assault on the idols of the mind can be read precisely in this register. His diagnosis of error is not moralistic but structural, aimed at forces operating beneath conscious awareness: linguistic habits, institutional pressures, and inherited authorities that distort perception itself. By naming and methodically countering these forces, Bacon sought to effect a transformation that exceeded his own person, one that would reorganize how humanity thinks, observes, and understands. In Neumann's sense, Bacon stands as a transpersonal agent of renewal, a figure through whom a culture attempts to move beyond the limits of its own unconscious inheritance.

Self-mythologizing in *The Wisdom of the Ancients*

Modern readers are often uneasy with the suggestion that a historical thinker might have understood himself in mythic terms. The reflex is to diagnose vanity, delusion, or rhetorical excess. Yet this discomfort is itself historically conditioned. In the early modern period, myth was not yet relegated to the nursery or the museum. It remained a serious mode of philosophical memory, a compressed language in which truths too deep or dangerous for direct statement could be preserved. Francis Bacon knew this, and *The Wisdom of the Ancients* should be read accordingly: not as a diverting curiosity, nor as a moral handbook, but as a work of symbolic self-situating.

To say that Bacon mythologizes himself is not to accuse him of fantasy. On the contrary, the myths he invokes are conspicuously tragic. They do not promise success, happiness, or recognition, but rather sacrifice and delayed vindication. Bacon does not imagine himself crowned; he imagines himself bound, expelled, or torn apart. This is not the mythology of narcissism, but of vocation; Bacon appears to have believed that certain roles recur in human history, and that he had been cast in one of them. *The Wisdom of the Ancients* is where he thinks this through most quietly and most honestly.

If this reading is accepted, then Bacon's reputation must be reconsidered. He was not merely a founder of modern science, nor merely a political operator who failed. He was a self-conscious archetypal figure, aware of the pattern he inhabited and willing to endure its cost.

Bacon's treatment of myth is strikingly unlike that of either medieval moralizers or Enlightenment skeptics. He does not reduce myth to ethical fable, nor does he dismiss it as primitive superstition. Instead, he treats myth as the symbolic residue of an early phase of human understanding, when philosophy, theology, poetry, and natural inquiry had not yet been forcibly separated. In this respect, Bacon stands closer to the Presocratic philosophers than to the scholastics who dominated the universities of his day.

For Bacon, myth encodes insight about nature, knowledge, and the human experience in imagistic form. It preserves what later systems forget. This is why *The Wisdom of the Ancients* sits so naturally alongside the *Advancement of Learning* and the *Novum Organum*. The latter works propose a new method; the former recovers an older wisdom that had been obscured by both dogma and disdain. Bacon does not see these projects as opposed. Recovery and innovation belong together.

This position places Bacon centuries ahead of the intellectual curve. Long before Vico, Herder, or Nietzsche, he recognizes that myth is not the enemy of reason, but one of its ancestral forms. Long before Jung, he understands that recurring symbolic figures represent stable patterns of human experience, what he himself, remarkably early, calls *archetypes*. The word is not incidental. It signals Bacon's belief that knowledge has deep structures, and that these structures manifest themselves both in nature and in narrative.

Bacon's introduction of *archetype* into English philosophical discourse is not a lexical curiosity; it is programmatic. By *archetype,* he means an original pattern, an exemplar that precedes and shapes its instances. Forms in nature, ideas in the mind, and figures in myth all participate in this logic of patterned recurrence. The implication is radical: human history, including intellectual history, unfolds according to intelligible shapes.

Once this is granted, the leap to self-recognition is not far. A thinker who believes that myth preserves archetypal patterns, and who believes himself engaged in a foundational act of intellectual renewal, could hardly avoid asking whether his own life participates in one of those patterns. *The Wisdom of the Ancients* can thus be read as Bacon quietly advancing that hypothesis.

Prometheus

Prometheus occupies a central place in Bacon's mythic constellation. Fire, for Bacon, is not merely warmth or technology; it is operative knowledge, the power to transform nature. Prometheus is therefore the archetype of the scientific benefactor, the one who advances humanity at personal cost.

Bacon's identification here is difficult to miss. His project aims at nothing less than the re-founding of human knowledge and power over nature. Yet he repeatedly emphasizes the dangers of this power, insisting on discipline, humility, and ethical restraint. Prometheus is punished not because knowledge is evil, but because it is perilous. Bacon's own fall from political power, following the publication of his most ambitious philosophical work, follows the same grim logic. The benefactor is bound; the gift survives.

Pan

In Pan, Bacon finds an image of nature itself, variegated, elusive, resistant to simplification. Pan inspires panic precisely because he cannot be grasped by abstract system. This figure maps neatly onto Bacon's hostility to premature theory and closed cosmologies. Nature must be pursued patiently, experimentally, and without arrogance. Biographically, Pan also explains Bacon's political difficulty. Courts favor clarity, loyalty, and fixed identities. Bacon's mind is restless, exploratory, and allergic to finality. Like Pan, he belongs to the margins rather than the center, to the woods rather than the palace.

Pallas Athena

Pallas Athena represents wisdom disciplined by law. Born from the head of Zeus, she symbolizes mind generating mind, knowledge emerging from intellect rather than tradition. Bacon's ambition is Athenaic rather than Promethean in temperament. He does not claim to have completed the sciences; he claims to have given them their constitution.

This is the logic of the Great Instauration. Bacon casts himself not as a discoverer of facts but as a founder of method, a lawgiver whose task is to establish the conditions under which discovery may proceed. Like Athena, his wisdom is civic, public, and ordered. And like many lawgivers before him, he is rejected by the city he sought to serve.

Orpheus

Orpheus embodies the civilizing power of learning, its capacity to soften manners, reconcile conflict, and draw even the inanimate into harmony. Bacon's hope that reform could occur without violence aligns closely with this figure. Knowledge should persuade, not coerce.

Yet Orpheus is torn apart by those he sought to civilize. Bacon suffers a parallel fate. His philosophy is celebrated; his character is dismembered. The work is detached from the life. Like Orpheus's singing head, Bacon's voice continues after death, while the body of his experience is left behind.

Oedipus

The most revealing identification is Oedipus. Bacon explicitly associates the Sphinx with science itself, nature posing riddles that destroy those who approach her wrongly. Oedipus succeeds not through force or revelation, but through interpretive intelligence. This is Bacon's ideal knower.

But Oedipus's reward is catastrophe. Truth brings knowledge of guilt, exile, and suffering. The bodily mark matters: Oedipus means "swollen foot." Bacon's lifelong affliction with gout in his foot, often incapacitating, carries an uncanny symbolic resonance. The wounded foot is the mark of the hero who cannot walk the ordinary path, who advances limping toward insight. Hampered physically, he is compensated mentally.

Oedipus saves the city and is then expelled to cleanse it. Bacon delivers a new law of knowledge and is cast out. The parallel is exact enough to suggest recognition rather than coincidence, and it also fits remarkably well with *Hamlet*. The Hamlet–Oedipus–Bacon parallel turns on a very specific kind of heroism: not the hero who acts, but the hero who understands. In Hamlet, as in Oedipus, the decisive act is not violence or conquest but interpretation. Hamlet reads signs: words, gestures, silences, the play within the play. Oedipus reads the riddle of the Sphinx. In both cases, intelligence triumphs where brute force would fail. This is precisely Bacon's ideal knower, the mind trained to interrogate appearances rather than submit to them.

But the reward is devastating. Truth does not enthrone either figure; it undoes them. Hamlet gains certainty at the cost of life; Oedipus gains knowledge at the cost of home, sight, and kingship. Yet in both cases the kingdom is preserved. Denmark is purged; Thebes is cleansed. The hero absorbs the catastrophe so that the polity may survive.

That pattern maps uncannily onto Bacon's self-understanding. He does not imagine the knower as a triumphant ruler. He imagines him as a sacrificial intelligence, one who bears the consequences of truth so that order can continue. This is why Oedipus matters so much. Bacon's association of the Sphinx with science is not incidental. Science is a riddle that destroys those who answer rashly, and even the one who answers correctly does not escape unscathed. The method succeeds; the man pays.

Seen this way, Bacon's project is neither naïvely optimistic nor coldly technocratic. It is tragic in the classical sense. Knowledge advances, but the knower is marked. Truth is preserved, but the hero is undone. That is the through-line from Oedipus to *Hamlet*—and it is exactly the role Bacon seems to have recognized, and accepted, for himself. This also helps explain why

Bacon could sustain such long-range confidence in posterity while enduring personal ruin in the present. Like Oedipus and Hamlet, he appears to have believed that truth outlives the hero, and that this, finally, is the only victory that matters.

Actaeon

In Actaeon and Pentheus, Bacon identifies a darker corollary to the civilizing figures of Orpheus and Prometheus: the fatal danger of knowing too much, of seeing what must not be seen. Actaeon is torn apart for glimpsing Diana unveiled; Pentheus is dismembered for spying on Dionysian rites forbidden to kings. Bacon reads these myths as political as well as moral warnings. They encode the peril of knowing the secrets of princes, the inner workings of sovereignty, the naked truth behind majesty. Such knowledge is not rewarded with enlightenment, but punished with obliteration. The resonance with Bacon's own position at court is unmistakable. As a counsellor who moved close to the hidden mechanics of power, he understood that insight itself could become a capital offense. The Actaeon figure is especially charged in the English context, where Elizabeth I was repeatedly figured as Diana. The lesson is severe and consistent: to look too closely upon the mystery of rule is to risk being torn apart by the very forces one sought to understand. Bacon's inclusion of these myths suggests not only historical insight, but personal knowledge. He knew that proximity to power, like proximity to truth, exacts its own brutal price.

Bacon as a Presocratic figure

Seen in this light, Bacon resembles a Presocratic philosopher more than a modern scientist. He stands at the threshold between myth and method, symbolic memory and experimental procedure. He looks backward to recover what was lost, and forward to establish what has never yet existed. This liminal position explains both his originality and his vulnerability. Transitional figures are rarely comfortable, and never secure. They belong fully to neither world. Bacon seems to have understood this, and to have accepted it.

He was centuries ahead of his time in his appreciation of the Presocratics; in the *Novum Organum* he wrote "They have got something of natural philosophy in them; they smack of the nature of things, of experience and of bodies." Elsewhere, in a tract unpublished during his lifetime, *Refutation of Philosophies*, he speaks thusly—

> It may perhaps be that you would like to hear my opinion of those other philosophers known to us, not in their own writings but through the writings of others—Pythagoras, Empedocles, Heraclitus, Anaxagoras, Democritus, Parmenides, etc. Here, my sons, I shall keep nothing back, but frankly open up to you the whole of my thought. Know then that I have, with the utmost zeal and patience, sought out the slightest breath of tradition about the findings and opinions of these men. Aristotle confutes them. Plato and Cicero quote them. Plutarch devoted an essay to them. Laertius wrote their lives. The poet Lucretius sings of them. All these sources, together with various other fragments and references which can be traced, I have sought out and read. Nor have I accorded them a contemptuous glance, but weighed them with patient fidelity... Inevitably some are better in one point, some in another. But if they be compared with Aristotle, my firm conviction is that several of them penetrated more shrewdly and deeply into nature in many points than he. It was inevitable that they should do so, since they were more devout devotees of experience than he. Especially is this true of Democritus, who by reason of his wide experience of the world of nature was even reputed to be a mage.

2. The hero is born to a royal virgin

> There be some whose lives are as if they perpetually played a part upon a stage, disguised to all others, open only to themselves.
>
> —Francis Bacon, "Of Friendship"

History remembers Elizabeth I as the Virgin Queen, serenely detached from erotic entanglement and ruling through symbolic chastity. Diplomatic correspondence from the first years of her reign tells a markedly different story. Between 1559 and 1561, ambassadors from Spain, Venice, France, and England itself repeatedly reported an intimacy between Elizabeth and Robert Dudley so conspicuous that it generated rumors of marriage, pregnancy, and even homicide across Europe. These reports were not pamphlets or polemics, but private intelligence sent by senior diplomats to their sovereigns and councils, preserved today in the great editorial series known as the *Calendar of State Papers*, collections that summarize and translate original dispatches housed in archives at Simancas, Venice, Paris, and London.

Early testimony comes from Count de Feria, Philip II's ambassador during the final months of Mary I and the opening of Elizabeth's reign. Writing privately to Spain in April 1559, de Feria reported that Elizabeth's attachment to Dudley was already notorious at court. In a dispatch dated 18 April 1559, he stated bluntly that "it is even said that her Majesty visits him in his chamber day and night." De Feria was Catholic and hostile to Elizabeth, but he was a seasoned grandee, writing confidentially to his king, with no reason to invent behavior that could easily be contradicted by other diplomats in London.

This assessment was independently confirmed by Venice. Giovanni Michiel Schifanoya, reporting to the Venetian Senate on 10 May 1559, described Dudley as enjoying extraordinary favor and familiarity with the Queen. In the *Calendar of State Papers Venetian*, his words are rendered to the effect that "My Lord Robert is in very great favour and very intimate with Her Majesty." Venice had no stake in English succession politics and no reason to echo Spanish hostility; Schifanoya's report demonstrates that Elizabeth's conduct was being noticed by neutral observers almost immediately after her accession.

From Paris, similar intelligence circulated through diplomatic networks. Giacomo Surian, writing in 1559, informed Venice that the Queen's affection for Dudley appeared so decisive that foreign courts believed it would determine her marital future entirely. His dispatch is commonly translated as stating that "the love which Her Majesty bears to Milord Robert is so great that she will end by marrying him, or else marry no one." Surian was not resident in London, but the significance of his report lies in how quickly the Dudley–Elizabeth relationship was being interpreted abroad as exclusive and potentially permanent.

Robert Dudley

A rumor that Elizabeth was pregnant bruited among the public; in August of 1560, one Anne Dowe of Brentwood, a sixty-eight-year-old widow, was jailed for speaking thus indiscreetly. Álvaro de la Quadra, Bishop of Aquila and successor to de Feria as Spain's representative, reported that a trusted informant had told him Dudley had sent to poison his wife, adding that Elizabeth's handling of marriage negotiations seemed designed to distract opposition until, as he put it, "this wicked deed" was accomplished. In early September, de Quadra met with William Cecil, Elizabeth's chief counselor, and wrote of the encounter—

> [Cecil] said that the Queen was going on so strangely that he was about to withdraw from her service . . . Lord Robert had made himself master of the business of the state and of the person of the Queen, to the extreme injury of the realm, with the intention of marrying her, and she herself was shutting herself up in the palace to the peril of her health and life. That the realm would tolerate the marriage, he said he did not believe . . . Last of all, he said that they were thinking of destroying Lord Robert's wife. They had given out that she was ill, but she was not ill at all; she was very well and taking care not to be poisoned . . . Since writing the above, I hear the Queen has published the death of Robert's wife.[5]

By the end of 1560, rumors of secret marriage and pregnancy were circulating so widely that even England's own representatives were forced to address them. Nicholas Throckmorton, Elizabeth's ambassador at the French court, wrote to William Cecil on 31 December 1560 that "the bruits of her doings be very strange in all courts and countries." He reported that the Spanish ambassador in France had asked him directly whether the Queen was not already secretly married to Lord Robert—an inquiry Throckmorton neither confirmed nor dismissed, but which demonstrates how far such suspicions had spread.

De Quadra's letters of January 1561 show that the crisis had not abated. Writing to Philip II on 22 January, he warned that if Elizabeth married Dudley without Spanish sanction, Philip had only to "give a hint" to her subjects and she would lose her throne. He described the Queen as "infatuated to a degree which would be a notable fault in any woman, much

[5] Letter to the Duchess of Parma, dated 11 September 1560

more in one of her exalted rank," and reported that Dudley himself claimed a willingness to restore Catholicism if Philip would countenance the match. In the same letter, de Quadra added cautiously, "Some say she is a mother already, but this I do not believe," a line that neatly captures both the intensity of the rumor and the diplomat's own restraint.

William Cecil, Francis Bacon's uncle and Elizabeth's chief advisor

Taken together, these dispatches form a coherent chronological record. From early 1559 through early 1561, multiple ambassadors independently observed or reported an intimacy between Elizabeth and Dudley that they believed had political consequences. The importance of this correspondence does not lie in proving marriage, pregnancy, or murder as historical facts, but in demonstrating how Elizabeth's private conduct was perceived by contemporaries, and how urgently foreign powers believed it threatened the stability of the English succession.

The reason Elizabeth did not marry Robert Dudley was not lack of affection, but the political reality that such a marriage threatened the stability of her reign. Diplomatic correspondence makes clear that Dudley was widely detested, and for reasons that contemporaries understood perfectly well. He was the son of the Duke of Northumberland, the architect of the 1553 coup that placed Lady Jane Grey on the throne, and the Dudley name remained indelibly associated with treason, executions, and the near-collapse of the Tudor succession. The mysterious death of Dudley's wife in 1560 made him politically radioactive; even unproven suspicion was sufficient to alienate nobles already predisposed to distrust him.

Beyond personal scandal lay deeper structural fears. A Dudley marriage would have meant the sudden dominance of a single court faction, the exclusion of rival noble houses, and a return to the hated politics of the royal favorite. Most dangerous of all was the succession question. A Dudley heir would inevitably be contested, potentially illegitimate in the eyes of large sections of the political nation, and capable of triggering the very civil war Elizabeth's accession had narrowly avoided. When de Quadra warned Philip II in January 1561 that a Dudley marriage could cost Elizabeth her throne with only "a hint" to her subjects, he was not indulging in hyperbole. He was articulating a consensus shared across courts and councils alike.

The virgin mother in mythology

In *The Origins and History of Consciousness,* Erich Neumann describes the virgin mother as a liminal figure, standing between collapsing authority and emergent order:

> The virgin mother, connected directly with the god who engenders the new order, but only indirectly with the husband, gives birth to the hero who is destined to bring that new order into being and destroy the old ... The hero's descent from the reigning family is symbolic of the struggle for the system of rulership, for that is what the struggle is really about.

This passage resonates uncannily with the Elizabethan situation. Elizabeth's refusal to submit her body to a husband, English or foreign, was not a withdrawal from power, but a rechanneling of it. The "reigning family" here is not merely dynastic; it is the exhausted Tudor settlement itself, strained by multiple succession crises, factionalism, and the unresolved trauma of civil war. In Neumann's terms, the virgin mother stands at precisely the moment when rulership must be reimagined rather than inherited in the ordinary way. Neumann goes further, emphasizing that the hero who emerges from such a matrix is almost always symbolically fatherless:

> The essence of the mythological canon of the hero-redeemer is that he is fatherless or motherless, that one of the parents is often divine ... mythology represents the hero as having two fathers: a personal father who does not count ... and a heavenly father who is the father of the heroic part, of the higher man, who is 'extraordinary' and immortal.

What matters here is not literal genealogy, but legitimacy displaced upward. Authority is no longer grounded in an ordinary paternal line, but in a higher, impersonal source: God, destiny, history, or, in Elizabeth's case, the body politic itself. When Elizabeth declared herself "married to the realm," she was performing precisely this displacement.

Neumann is explicit that the term *virgin* must not be misunderstood in modern or moralistic terms:

> These mothers are virgin mothers, which is not to say that what psychoanalysis has attempted to read into this fact is necessarily correct... virginity simply means not belonging to any man personally; virginity is in essence sacred, not because it is a state of physical inviolateness, but because it is a state of psychic openness to God.

This definition is crucial. Elizabeth's virginity was not primarily anatomical; it was juridical and symbolic. She belonged to no man, no faction, no foreign prince. That "psychic openness" Neumann describes maps cleanly onto Elizabeth's insistence that her authority derived directly from God and from England as a corporate body, not from a husband who might subsume or redirect it.

The pregnancy portrait at Hampton Court

Among the more unsettling images from Elizabeth's reign is an allegorical pregnancy portrait now at Hampton Court Palace. The Queen appears in the guise of Diana, accompanied by a stag. Elizabeth was repeatedly identified with Diana, the virgin goddess of chastity and sovereign power over nature. Court poets and artists returned to this figure frequently, for Diana allowed Elizabeth's unmarried state to be transfigured into divine authority.

Pregnancy portrait of Elizabeth I at Hampton Court Palace

The presence of the stag complicates the image. The stag unmistakably invokes the myth of Actaeon, the hunter who accidentally glimpses Diana naked and is punished by being transformed into a stag, torn apart by his own hounds. He represents the man who sees too much, who intrudes upon forbidden knowledge, who crosses the invisible line between reverence and presumption. To pair Elizabeth-as-Diana with Actaeon is to encode a warning: there are things at court that must not be looked at directly, questions that cannot be asked without consequence.

The portrait's unease is heightened by its history. Like many images of Elizabeth, it shows signs of reworking and overpainting. Details have been softened or obscured, though no surviving evidence allows the original composition to be reconstructed with certainty. What remains is an image saturated with anxiety.

This logic, power maintained through controlled visibility, finds a striking parallel in Francis Bacon's treatment of the Actaeon myth in *Wisdom of the Ancients*. Bacon does not read Actaeon as a moral tale about prurience; he reads it as a parable of political danger. Actaeon, he suggests, represents the courtier or statesman who hunts after the secrets of princes, who grows too curious about hidden counsels. Once transformed into a stag, such a man becomes exposed and vulnerable, no longer master of events but prey to them. The hounds that destroy him are not external enemies, but his own thoughts, anxieties, and disclosures, which turn upon him once he has crossed the fatal threshold of knowledge.

In Bacon's reading, Actaeon is destroyed not for moral corruption but for epistemic transgression. He sees what cannot be safely seen and knows what cannot be safely known. The myth reveals a world in which truth is not denied, but displaced, forced into allegory, where it can be contemplated indirectly without being spoken. Power depends not merely on secrecy, but on the careful staging of secrecy: on making the boundary visible while keeping what lies beyond it hidden. This insight is not incidental to Bacon's life. It is the key to understanding it.

Hamlet and Queen Elizabeth

Otto Rank mentions *Hamlet* as an example of the heroic archetype: "The fable of Shakespeare's *Hamlet* also permits of a similar interpretation… mythological investigators bring the Hamlet legend from entirely different viewpoints into the correlation of the circle of myths." Rank's remarks on *Hamlet* are of interest:

> It seems to me not improbable that the inspired poet portrayed himself in the Danish prince, so that he might with impunity utter high treason . . . the participation of Hamlet in his entrapping play might be explained from the fact that powerful opponents of Elizabeth did really use the poet as a means to attack her and stir her conscience. In this case, we should have a reflection, in Hamlet's editing of the "play," of the part important friends of the poet actually had in his work.[6]

Rank voiced doubts over the attribution of the canon: "we know so little of his actual life and even doubt his authorship. Shakespeare's work and the biographical material that has been gathered about the Stratford butcher's son have just as much psychological connection as have the Homeric poems and our scanty information about the blind Ionian singer."[7]

Hamlet enters print in the early Jacobean moment, immediately after the death of Elizabeth in 1603. During her lifetime, Elizabeth was not merely a monarch but a sacralized political symbol: the Virgin Queen, the body politic itself, the "Great Mother" of England. To stage a tragedy in which the queen-mother is morally compromised, sexually suspect, and implicated in the corruption of the realm would have been extraordinarily sensitive while Elizabeth lived. After her death, that symbolic restraint loosens.

Read in that light, Gertrude's sudden centrality in *Hamlet* is striking. She is not a passive background figure; the moral problem of the play runs through her body. Hamlet's revulsion is not primarily directed at Claudius's crime (regicide is almost taken for granted in revenge tragedy), but at Gertrude's remarriage, her sexuality, her apparent lack of discernment, her failure to mourn "with the constancy of nature." This aligns uncannily with long-suppressed cultural anxieties surrounding Elizabeth: her refusal to marry, the obsessive scrutiny of her body, rumors of illicit intimacy (Leicester, Essex), and the sense that the fate of the realm was bound to the queen's sexual and maternal status. Gertrude becomes, in effect, the dramatized site where those anxieties can finally be spoken.

---

[6] Rank, Otto. *The Myth of the Birth of the Hero and Other Writings*. New York: Vintage 1959 p. 237

[7] Ibid., p. 199

What makes the parallel sharper is that *Hamlet* is not an allegory of succession politics in the abstract; it is a meditation on a realm poisoned at its maternal source. Denmark is "an unweeded garden" not simply because a king was murdered, but because the queen's desire has short-circuited lawful order. That emphasis resonates with post-Elizabethan retrospection: once the Virgin Queen is gone, her symbolic immunity evaporates, and the culture can begin to re-imagine the maternal sovereign as fallible, opaque, even culpable. Gertrude is not Elizabeth *as she was*, but Elizabeth *as she could now be thought*, no longer protected by living majesty.

This also explains why Hamlet is so uniquely modern: he is the son of a mythic political mother, suddenly forced to see her as human, sexual, compromised. That psychic rupture mirrors England's own transition from the long Elizabethan stasis into the uncertain Jacobean present. The play's appearance *after* Elizabeth's death is therefore not merely chronological convenience; it marks a release. The maternal body of the state can finally be interrogated, and the cost of that interrogation—melancholy, paralysis, moral hyper-consciousness—is dramatized in Hamlet himself.

Within this framework, Polonius stands out as the clearest historical analogue. His resemblance to William Cecil, Elizabeth's principal counselor for four decades, is widely acknowledged, even within orthodox scholarship. Polonius's habits of surveillance, his prolix moralizing, his use of children as political instruments, and his belief that stability is preserved through information control closely mirror Cecil's governing style and reputation. Gertrude's relation to Elizabeth, by contrast, is less biographical than symbolic. She represents Elizabeth not as a person, but as a problem of sovereignty: a queen whose body, marriage, and sexuality are inseparable from the fate of the state.

Although *Hamlet* enters the historical record in print only in 1603, the evidence is overwhelming that it circulated in manuscript for several years before the death of Elizabeth. The so-called "Bad Quarto" of 1603 (Q1), far from being an embarrassment, is the clearest proof of this pre-1603 life. Its shortened length, paraphrased speeches, rearranged scenes, and simplified language are characteristic of memorial reconstruction, most plausibly by an actor who had performed the play repeatedly. A text of this kind presupposes an already established performance history: lines spoken often enough to be half-remembered, scenes vivid enough to be recalled in outline, but not yet fixed in authoritative literary form. In other words, *Hamlet* did not suddenly appear after Elizabeth's death; it had already been living on the stage during her final years, circulating orally and theatrically rather than textually.

This chronology is crucial for interpretation. During Elizabeth's lifetime, the play seems to have existed in a state of deliberate instability, performed publicly but not fully released into print. Only after her death in March 1603 does *Hamlet* break into textual circulation, first in damaged form (Q1), then in the expansive and philosophically dense second quarto of 1604–05. Such a sequence strongly suggests political and symbolic constraint. A tragedy centrally preoccupied with a morally compromised queen-mother, her sexuality, her remarriage, and her implication in the sickness of the body politic would have been extraordinarily sensitive while the Virgin Queen herself still embodied the state. The play could be performed, absorbed as drama, but not yet fixed, authorized, or stabilized as text. Once Elizabeth was gone, that restraint loosened. The maternal figure at the heart of the realm could finally be interrogated, and *Hamlet* could emerge fully into print. This historical pattern strengthens the case for reading Gertrude not as a casual character flaw, but as a dramatized site where long-suppressed anxieties about Elizabeth's body, authority, and symbolic motherhood could at last be explored.

3. The hero is raised by foster-parents

> "It is not my meaning to treat him as a ward: Such a word is far from my Motherly feeling for him. I mean to do him good."
>
> —Lady Anne Bacon in a letter to Anthony Bacon, 18 April 1593

Francis Bacon was raised in a household that stood at the nerve center of Elizabethan governance, learning, and reform. His father, Sir Nicholas Bacon, served as Lord Keeper of the Great Seal and was widely respected for his legal prudence, moderation, and commitment to institutional stability. Though not a grand stylist or visionary theorist, Sir Nicholas embodied the virtues of the Tudor civil servant: administrative competence, fidelity to the crown, and a belief that law was the principal instrument through which order could be maintained in a fragile post-Reformation polity. His career placed Francis, from infancy, within sight of the highest legal and political mechanisms of the realm.

Lady Anne Bacon, by contrast, brought into the household a formidable humanist and theological intensity. Anne Bacon, née Cooke, was among the most learned women in England; her father, Sir Anthony Cooke, had tutored Edward VI. Fluent in Latin, Greek, Italian, and French, she translated major Protestant theological works and maintained correspondence with leading reformers on the Continent. Her piety was rigorous, introspective, and morally exacting, shaped by the evangelical wing of the English Reformation. If Sir Nicholas represented the stabilizing force of law and office, Lady Anne represented the claims of conscience, scripture, and inward discipline. Francis thus grew up in a household where legal reason and moral seriousness were not opposed, but held in constant dialogue.

This domestic world was further intensified by its proximity to the household of William Cecil, Elizabeth I's principal minister and the architect of much of her long and successful reign. Cecil was married to Lady Anne Bacon's sister Mildred, herself a renowned scholar and patron of learning. Mildred Cecil shared her sister's humanist education and religious seriousness, and the Cecil household became an extension of the Bacon domestic sphere, a kind of informal academy where politics, theology, classical learning, and statecraft intersected daily. The young

Francis would have encountered there not only the rhetoric of power but its burdens: the necessity of prudence, concealment, patience, and compromise in service of the commonwealth.

Bacon's childhood home, Gorhambury House
near St. Albans, Hertfordshire

The Cecil household, however, also exemplified a particular tension that would later shape Bacon's thought. William Cecil was cautious, incremental, and deeply suspicious of novelty; his governing philosophy prized continuity and balance over systemic reform. Lady Anne Bacon's moral absolutism, by contrast, could verge on severity, even toward her own sons. Francis grew up, therefore, between two powerful models of authority: one grounded in tradition and political survival, the other in inward truth and reformist zeal. His later ambition, to reform knowledge itself while remaining loyal to the state, can be read as an attempt to reconcile these inherited imperatives.

In this sense, Bacon's early formation was not merely privileged but structurally charged. He was raised at the crossroads of law, religion, and governance, within families that effectively *were* the Elizabethan regime. The ideals and limitations of that world, its reverence for order and its fear of disorder, its moral earnestness and its resistance to radical change, formed the matrix against which Bacon would later define his own project. His call for a "Great Instauration" did not arise in opposition to the Tudor state, but from within its most intimate households, shaped by the very people who had built and sustained it.

If Elizabeth's virginity functioned as a political technology, Francis Bacon's instauration functions as its epistemic counterpart: a deliberate attempt to remake the foundations of knowledge itself. Long before the *Novum Organum* (1620) formally announced the overthrow of Aristotle's *Organon*, Bacon conceived what he would later call the *Great Instauration*: a total reform of learning, method, and the uses of human understanding. As Macaulay observed, "His gigantic scheme of philosophical reform is said by some writers to have been planned before he was fifteen, and was undoubtedly planned while he was still young."

In terms of heroic myth, this precocity matters. In Lord Raglan's canonical pattern, the hero's decisive act is a victory over the king, a giant, dragon, or wild beast, a symbolic defeat of an overgrown, tyrannical power. For Bacon, that adversary would be Aristotle himself: the senex ruler of the intellect, whose authority had ossified into scholastic domination. Raglan also notes, almost as a rule, that we are told nothing of the hero's childhood. Here the pattern fits imperfectly but suggestively. Bacon's early years are not entirely unknown, yet the surviving record is strikingly thin, especially in the treatment of his childhood by his great Victorian editor and biographer James Spedding. Spedding is notoriously brief on the formative interior life of the boy; what we possess are outlines rather than narratives, enough to suggest early exceptionalism, but not enough to domesticate it. The effect is archetypal: Bacon appears early as a mind already oriented beyond ordinary development, with no richly detailed childhood to anchor him securely within the common run of men.

Miniature of Francis Bacon at seventeen by Nicholas Hilliard

One emblem of this precocity is the famous miniature by Nicholas Hilliard, painted when Bacon was seventeen or eighteen. Around the oval runs a Latin inscription usually rendered: *Si tabula daretur, animum mallem*—"If only I could paint his mind." The phrase is extraordinary. It presents the young Bacon not as a court ornament or dynastic heir, but as an intellect already exceeding representation. In heroic terms, it marks him as inwardly "other," defined less by lineage than by a latent, almost dangerous capacity.

That capacity was cultivated early. Bacon was sent to Trinity College, Cambridge, at the age of twelve, together with his elder brother Anthony Bacon, who would later play a crucial role in European intelligence networks. In the 1665 book *Statesmen and Favourites of England since the*

Reformation, David Lloyd said of Francis that "At twelve his industry was above the capacity and his mind beyond the reach of his contemporaries." Macaulay's phrase, that Bacon possessed "the most exquisitely constructed intellect that has ever been bestowed upon any of the children of men" suggests something more specific and more structural than raw talent alone. An "exquisitely constructed" intellect is not merely powerful; it is formed, shaped, assembled with unusual care. The metaphor implies architecture rather than accident, design rather than spontaneity.

What formed Bacon was not simply "a good education," nor even extraordinary intelligence, but a configuration of pressures, intellectual, moral, political, and psychological, that is no longer available, and in many respects should not be. Bacon was shaped in a world that placed enormous demands on children of the governing elite. Expectation was relentless; moral scrutiny was intense; failure carried real consequences not only for oneself but for one's family and nation. He absorbed law, theology, rhetoric, and politics, not as optional attainments but as matters of survival and duty. The early modern household, especially one so close to power, did not distinguish sharply between childhood and preparation for adult responsibility. Formation began early because it had to.

Cambridge proved decisive not because it satisfied Bacon, but because it disappointed him. There, still a child, he encountered Aristotelian scholasticism in its late, arid form; a system he would later describe as having "done more harm than good," and whose methods he believed actively obstructed the advancement of knowledge. The seeds of the future rebellion were planted not in ignorance, but in premature overexposure to an authority he already judged inadequate.

When Bacon went up to Cambridge, he entered not simply a university but a contested intellectual arena, one in which humanist learning, scholastic tradition, and emerging reformist energies coexisted uneasily. One of his early tutors was John Whitgift, later Archbishop of Canterbury, a figure emblematic of institutional discipline and doctrinal order. Whitgift was a rigorous logician and a staunch defender of ecclesiastical authority, deeply committed to the preservation of order against both Puritan radicalism and intellectual novelty. Under such tutors, Bacon would have received a demanding training in dialectic and disputation, precisely the methods he would later criticize as sterile when treated as ends in themselves. The experience seems to have sharpened his sense that education could be simultaneously exacting and misdirected: powerful in form, yet deficient in its orientation toward discovery.

More broadly, Bacon encountered at Cambridge a pedagogy still dominated by formal disputation and commentary, where Aristotle was not merely studied but institutionalized as an authority. Lectures were conducted in Latin; intellectual progress was measured by agility in syllogism, rather than by engagement with nature. Bacon's dissatisfaction with this regime, formed not in ignorance but in intimate exposure, was therefore experiential rather than abstract. He did not reject scholasticism from the outside; he endured it, mastered it, and found it wanting. The seeds of his later insistence on method, experiment, and the reform of learning were planted precisely in this environment, where intellectual energy was abundant but systematically misapplied.

Taken together, the presence of figures like Whitgift and Harvey underscores that Bacon's Cambridge years were not barren or merely repressive. They were formative in a more paradoxical sense. He was trained by men of formidable intellect and authority, exposed to new vocabularies of mind and self, and initiated into the institutional habits of learned culture. What emerged from this was not rebellion for its own sake, but a conviction that learning required re-founding at the level of its aims. Cambridge did not give Bacon his mature philosophy, but it gave him, early and decisively, the sense that the existing order of knowledge, however disciplined, had reached a point of exhaustion.

4. The hero returns to his kingdom

After leaving Cambridge in 1576, Bacon was sent to France as part of the English embassy under Amyas Paulet. This period, often treated as merely preparatory, is in fact crucial. Moving through the courts of France while still a teenager, Bacon absorbed continental politics, diplomacy, and most importantly, secrecy. It was during this time that he devised what he later called his biliteral cipher: a method of encoding a secondary message within an innocuous primary text by using two subtly different typefaces or letter forms, conventionally labeled "A" and "B." Any visible text could thus carry a hidden message readable only by those who knew the key.

Technically speaking, this was the first explicit binary system for encoding information: meaning reduced to a sequence of dual oppositions, long before modern computation. Symbolically, it is hard to imagine a more Baconian invention. Knowledge is no longer a matter of inherited authority or surface appearance, but of method, structure, and decoding. Even as a youth, Bacon was already practicing the logic of the *Novum Organum*: the replacement of verbal tradition with operational technique. In heroic-archetypal terms, the future dragon-slayer is already forging his weapon.

Seen this way, Bacon's *instauration* does not emerge suddenly in middle age. It unfolds as the life-long working-out of a heroic vocation formed under a virgin polity, directed against an ancient intellectual monarchy, and announced early through signs of precocity, displacement, and symbolic invention. The child is not fully visible, but the hero's task already is.

France offered a living demonstration of political authority under strain, played out before the eyes of Europe at Blois. The Estates-General of 1576–77, summoned by Henri III amid civil war and fiscal collapse, was the first such assembly since 1560–61, and it met not as a ceremonial relic but as a crisis parliament. Convened at the Château de Blois, it brought together clergy, nobility, and Third Estate to negotiate money, policy, and legitimacy at a moment when the crown could no longer command unquestioned obedience. The monarchy was not overthrown, nor constitutionally transformed, yet it was unmistakably exposed. Subsidies were refused, grievances formally aired, and royal authority subjected to bargaining in full public view. The Estates made visible what theory often obscures: that sovereignty, when stripped of inherited awe, depends on consent, performance, and persuasion.

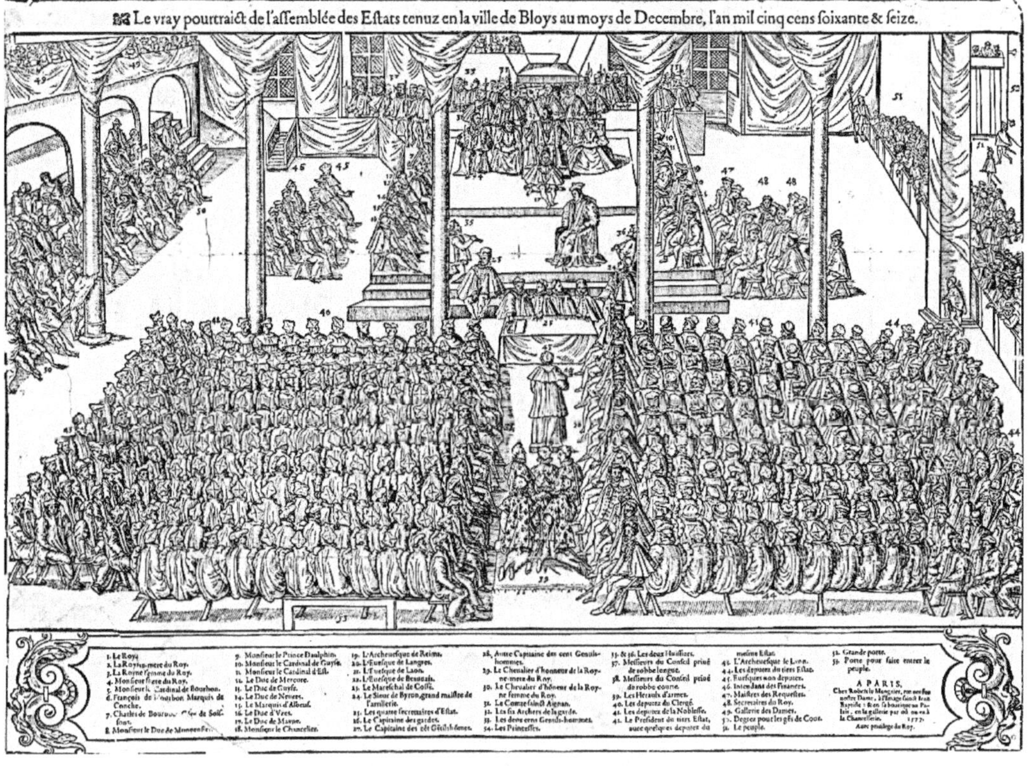

Estates-General of 1576-77

Bacon was present at exactly the moment when monarchy was compelled to negotiate with collective power, and he witnessed politics conducted not as abstraction but as theatre: procession, ritual, grievance, and refusal staged with deliberate solemnity. The lesson aligns uncannily with his later career. When he returned to England, he returned having seen both how authority conceals itself, and how, under sufficient strain, it can be made to speak.

Bacon's return to England in early 1579 was precipitated abruptly by the death of his father, Sir Nicholas Bacon. The event marked a decisive rupture. The elder Bacon's death removed the material and institutional security on which his future had quietly depended. Instead of returning as a courtier-in-training eased into place by paternal authority, Bacon returned as a young man of extraordinary intellectual ambition suddenly exposed to precarity. The mythic pattern sharpens here: the hero re-enters his native realm not in triumph, but under constraint, stripped of protection, compelled to make his way by talent rather than by inheritance.

The immediate consequences were sobering. Bacon inherited little, fell into debt, and found himself compelled to pursue advancement through the law, the Inns of Court, and parliamentary service rather than through effortless patronage. Yet this enforced recalibration proved formative. As Neumann writes,

> It is precisely the persecutions and dangers heaped upon him by the hateful father figure that make him a hero. The obstacles put in his way by the old patriarchal system become inner incentives to heroism, and, so far as the killing of the father is concerned, Rank is quite right when he says that "the heroism lies in overcoming the father, who instigated the hero's exposure and set him the tasks." It is equally right to say that the hero, "by solving the tasks which the father imposed with the intent to destroy him, develops from a dissatisfied son into a socially valuable reformer, and conqueror of man-eating monsters that ravage the countryside, an inventor, a founder of cities, and bringer of culture." But only if we take the transpersonal background into account do we arrive at an interpretation which does justice to the hero as a maker of human history, and which sees in the hero myth a great prototypal event honored by all mankind.

The lessons absorbed abroad, about secrecy, persuasion, institutional weakness, and the performative nature of power, now had to be translated into survival. Bacon's early legal training, his disciplined prose style, and his acute sensitivity to faction all belong to this period of compression, when intellect was forced into practical channels. What might otherwise have remained speculative became tactical.

From this point forward, the trajectory of Bacon's life acquires its distinctive double aspect. On the surface, he advances slowly and often painfully through public office: barrister, parliamentarian, counselor, eventually attorney general, lord keeper, and lord chancellor. Beneath that visible ascent runs a parallel project, more radical and more enduring: the reconstruction of knowledge itself. The frustrations of court life, the spectacle of authority under strain witnessed in France, and the early shock of loss at home converge into a single conviction: that inherited systems, whether political or intellectual, had exhausted their authority and required renewal at the level of method. Bacon's return, therefore, is not merely biographical. It is the moment when exile becomes vocation, and experience hardens into a lifelong program aimed not at power alone, but at the re-founding of how power knows, governs, and endures.

5. The hero is victorious

> For as the works of wisdom surpass in dignity and power the works of strength, so the labours of Orpheus surpass the labours of Hercules.
>
> —Francis Bacon, *Wisdom of the Ancients*

Across twentieth-century mythological theory, scholars repeatedly identified a striking and persistent pattern at the core of heroic narratives: the displacement, overthrow, or symbolic killing of a paternal authority who once guaranteed order but has become obstructive to life, growth, or truth. Otto Rank first articulated this structure in psycho-mythic terms, observing that the hero's defining act is not mere rebellion but separation from an overbearing father whose authority has hardened into fate. Lord Raglan subsequently stripped the motif of overt psychology and demonstrated its recurrence as a formal narrative sequence: the hero arises at a moment of sterility or decay, confronts a ruling figure whose legitimacy derives from custom rather than vitality, and inaugurates a renewed order without annihilating the cosmos the father once sustained. Joseph Campbell generalized this pattern into the monomyth, where the hero's task is not to destroy the world but to renew it by returning with a boon that frees the community from paralysis. Erich Neumann, writing from depth psychology, reframed the same structure as the ego's emancipation from the "Great Father," a necessary differentiation that allows consciousness to move from dependence on inherited forms toward autonomous engagement with reality.

Across these thinkers, one point is crucial: heroic "patricide" is rarely literal and almost never moralized. The father is not typically evil; he is exhausted, overextended, or bound to a prior epoch. His authority has become static, no longer responsive to the living world. The hero therefore performs an act of lawful succession, rather than nihilistic destruction. Order is preserved even as sovereignty changes hands. The mythic drama lies not in chaos but in transfer, in the movement of authority from past to future, from inheritance to generation, from repetition to renewal. When translated from political or familial narratives into intellectual history, this structure becomes especially revealing, because the "father" need not be a person at all. He may be a system, a canon, or an inherited epistemic regime.

By the late sixteenth century, Aristotle occupied precisely such a position. He was no longer merely a philosopher among others, but an institutional sovereign whose authority structured the entire apparatus of European learning. Scholastic curricula were organized around Aristotelian categories; natural philosophy proceeded primarily by commentary and deduction; and intellectual authority flowed in a rigid, hierarchical chain from Aristotle to commentators, from commentators to teachers, and from teachers to students. Innovation under this regime did not mean discovery but interpretation. To think was to gloss; to dissent was to misread. Francis Bacon's repeated language of idols, tyranny, servitude, chains, and enchantment is therefore not incidental rhetoric or polemical excess. It reflects his lived experience of Aristotelianism as a regime, one that governed not only conclusions but the very form of inquiry itself.

In this sense, Aristotle functioned as a "king" of philosophy, and in some contexts as a tyrant in the technical early modern sense of the term: an authority obeyed by custom rather than by continual justification. Bacon does not accuse Aristotle of intellectual fraud or moral corruption; instead, he diagnoses a historical pathology in which a once-fertile rule has hardened into unquestioned sovereignty. Tyranny here is not personal vice but temporal mismatch, rule persisting beyond its proper season. The world has changed; experience has accumulated; instruments have improved. Yet philosophy remains enthralled to a paternal authority whose categories were never designed to govern an expanding empirical universe.

The *Novum Organum* performs its intervention at exactly this mythic register. The title itself announces succession rather than reform. Aristotle's *Organon* had long been understood as the logical instrument of knowledge; Bacon's *new instrument* implicitly displaces the old without requiring its demonization. Bacon does not say, "Let us correct Aristotle." He says, in effect, "Let us begin again." That rhetorical stance is decisive. It mirrors heroic father-overthrow myths in which the old king is not wicked but sterile, no longer capable of sustaining growth. The land does not flourish under his rule, not because he has failed morally but because time has moved on.

Novum Organum Scientiarum (1620)

Bacon's induction as democratic epistemology

The year 1620 marks a striking convergence of intellectual and political transformation. In that year Francis Bacon published the *Novum Organum*, announcing a new instrument of knowledge grounded in experience rather than inherited authority, while across the Atlantic the Mayflower Compact articulated a radically new principle of political legitimacy: government arising from mutual consent, rather than divine right or dynastic inheritance. The parallel is not superficial. Both documents enact a decisive shift from top-down to bottom-up authority, from paternal command to collective process, from legitimacy grounded in precedent to legitimacy grounded in procedure. Bacon's overthrow of Aristotelian deduction and the colonists' eventual rejection of imposed sovereignty belong to the same deep cultural movement: the displacement of the Father as the unquestioned source of order.

In epistemological terms, Bacon's target was the paternal authority of Aristotle, whose *Organon*, based on top-down deduction, was the standard text on logic. Knowledge flowed downward from axioms to particulars, from master to student, from past to present. The *Novum Organum* (New Organon, "New Instrument") reverses this flow. It insists that knowledge must rise slowly from experience, tested collectively, corrigible by future inquiry, and oriented toward what Bacon repeatedly calls "the relief of man's estate." Authority is no longer vested in a person or a canon, but in a method, an impersonal, repeatable process that any properly disciplined mind may employ. In political terms, the Mayflower Compact performs an almost identical inversion. Authority no longer descends from king to subject by hereditary right; it arises from the consent of those who bind themselves together into a civil body politic. Sovereignty is relocated from the Father-King to the procedural agreement of the community.

Seen through the lens of Otto Rank, both events represent acts of symbolic patricide. Rank's hero must confront and displace an overbearing father whose authority structures the world but stifles growth. In Bacon's case, the father is Aristotle as institutional sovereign; in the colonial case, it is the Old World monarch whose legitimacy is inherited rather than renewed. Yet in neither instance is the father demonized. Aristotle is praised as a great intellect; the king remains theoretically honorable. What is rejected is not the father's person but his continued rule. Rank's pattern holds: the hero does not destroy the world the father made possible; he frees it from paternal fixation so that life may continue under new conditions.

Lord Raglan's royal hero pattern clarifies the political dimension. Raglan emphasizes that heroic succession often arises at moments of sterility, when the land no longer flourishes under the reigning order. Scholasticism had become intellectually barren; Europe's religious and political hierarchies were increasingly unable to govern expanding worlds, literal and conceptual. In Raglan's terms, both Aristotelian philosophy and Old World monarchy function as sterile kingships. The heroic response is not mere rebellion but lawful displacement. The *Novum Organum* does not abolish reason; it restores fertility to it through induction. The Mayflower Compact does not abolish order; it regenerates it through consent. In both cases, the heroic act clears space for continuity by ending a reign that can no longer sustain growth.

Erich Neumann's account of ego development provides the deepest psychological resonance. Neumann describes modernity as the painful but necessary separation of consciousness from the Great Father, authority experienced as absolute, protective, and suffocating. Bacon's method is precisely such a separation in intellectual life. Nature must no longer be known through paternal categories imposed in advance, but encountered directly, patiently, and humbly. Likewise, the political autonomy claimed by the colonists represents a collective ego-separation from the archaic Father of divine monarchy. The risk in both cases is dissolution: error, anarchy, chaos. The response is method. Bacon counters epistemic chaos with disciplined induction; the colonists counter political chaos with covenantal self-binding. Autonomy is not license but responsibility.

Joseph Campbell's monomyth allows these parallel revolutions to be seen as a shared heroic journey. The Old World is the known but exhausted realm; the New World, both geographical and epistemological, is the space of trial and danger. Bacon's *Novum Organum* functions as a "boon" brought back from the threshold—a method rather than a doctrine, a tool others must use. The Mayflower Compact similarly offers no substantive political theology, only a procedural framework by which a community may govern itself. In Campbell's terms, both are returns with the boon that enable collective renewal rather than individual glory. The hero in each case is not a conqueror but a founder of process.

Taken together, these 1620 revolutions suggest that modernity did not begin with a single act of defiance but with a coordinated reconfiguration of authority across domains. Knowledge and governance alike were re-grounded in bottom-up legitimacy, procedural validation, and future-oriented responsibility. Bacon's induction and the colonists' covenant are homologous structures: both displace the Father without abolishing order, both replace command with method, and both inaugurate worlds in which authority must continually justify itself through results rather than lineage. In mythic terms, 1620 marks not the death of the Father, but his successful succession.

6. The hero becomes king

Raglan's thirteenth point, *the hero becomes king,* finds a suggestive realization in Francis Bacon, though not in the conventional register of hereditary sovereignty. Bacon never wore a crown; yet few figures of the early modern period so consciously assumed kingship in another sense, mastery over the realm of knowledge itself. As a young man he wrote to his uncle William Cecil, Elizabeth's chief minister, that he had "taken all knowledge to be my province." The phrase is not casual rhetoric. It announces an imperial conception of the intellect, a claim to jurisdiction rather than mere curiosity. Knowledge is not simply something to be possessed or admired; it is territory to be surveyed, ordered, governed, and made productive for the commonwealth. In this sense Bacon casts himself as an epistemological king, a ruler whose dominion is the mind and whose laws are methods.

This symbolic kingship briefly approached political embodiment in 1617, when James I of England traveled to Scotland for the anniversary celebrations of his accession. During the king's absence, Bacon, as Lord Chancellor, effectively presided over the highest legal authority in the realm. Seated at the Court of Chancery in Westminster Hall, he occupied the chair that mediated between royal prerogative and equitable justice. A contemporary account notes that Bacon appeared robed head to toe in purple, the color reserved by sumptuary law for royalty and the highest ranks of nobility, the quantity and richness of the dye itself a visible index of authority. He had worn the same color at his wedding, an event staged with unmistakable ceremonial ambition. These details matter not as gossip, but as symbolic gestures: Bacon understood power as theatrical, juridical, and visible, and he knew how to inhabit its forms.

That understanding was already marked in the earliest biographical account devoted to him, found in *Histoire naturelle de Mre. Francois Bacon* (1631) by Pierre Amboise, which declares that Bacon was "born in the purple and brought up with the expectation of a great career." Whether meant literally, metaphorically, or diplomatically, the phrase crystallizes the impression Bacon left on contemporaries: a man marked for rule. Read alongside Bacon's own claim to universal intellectual jurisdiction, the line acquires a deeper resonance. He is king not because he inherits a throne, but because he founds a sovereignty, one grounded in law, method, and disciplined inquiry. In Raglan's terms, the pattern holds: the hero becomes king. But in Bacon's case, the kingdom is knowledge itself, and the crown is worn by the mind.

7. The hero prescribes laws

> All those which have written of laws, have written either as philosophers or as lawyers, and never as statesmen. As for the philosophers, they make imaginary laws for imaginary commonwealths, and their discourses are as the stars, which give little light because they are so high. For the lawyers, they write according to the states where they live, what is received law, and not what ought to be law; for the wisdom of a law-maker is one and of a lawyer is another.
>
> —*Advancement of Learning*

In heroic myth, lawgiving is not bureaucratic legislation, but a cosmological act: the hero articulates the principles by which chaos is restrained and the future made possible. In this sense, Bacon stands closer to Solon or Lycurgus than to any ordinary jurist. His legal reforms, his moral essays, and above all his attempt to legislate the conduct of the human intellect, together form a single lawgiving gesture, one aimed not merely at England, but at civilization itself. In a sense, they are of a piece:

> Bacon's procedures in natural philosophy were adapted by him from procedures in English law. Specifically, certain procedures in legal investigation and court trials, when linked with Bacon's own proposals for legal reform, not only exactly parallel his procedure for a reformed natural philosophy, but they were the model for it.[8]

Francis Bacon's role as a lawgiver begins not in metaphor, but in practice. Trained at Gray's Inn and rising through the highest legal offices of the realm—Solicitor General, Attorney General, and finally Lord Chancellor—Bacon approached the law not as a mere technician but as a reformer concerned with foundations, coherence, and moral purpose. His legal writings, many of them circulated in manuscript rather than print, repeatedly seek to clarify, systematize, and rationalize the English common law, stripping it of obscurity, redundancy, and arbitrary custom. In this respect Bacon stands closer to the classical lawgiver than to the routine jurist: his ambition was not simply to argue cases, but to understand law as an ordered system capable of improvement. As Daniel R. Coquillette

---

[8] Julian Martin, *Francis Bacon, the State, and the Reform of Natural Philosophy* 1992, Cambridge University Press

observes, "Bacon was the first truly analytical and critical jurist in the Anglo-American tradition. He was also the first to dare to apply an empirical, inductive analysis to lawmaking." This judgment captures Bacon's distinctive position: he brought to law the same reforming intelligence he would later bring to natural philosophy, treating legal institutions as human constructs subject to analysis, testing, and correction.

> "Laws are made to guard the rights of the people, not to feed the lawyers. The laws should be read by all, known to all. Put them into shape, inform them with philosophy, reduce them in bulk, give them into every man's hand."
>
> —Francis Bacon, speech in the House of Commons on 26th February 1593

Bacon's tenure as Lord Chancellor reinforces this heroic-lawgiver profile. Contrary to later caricatures of the Court of Chancery as dilatory and labyrinthine, Bacon sought to accelerate equity proceedings, regularize procedure, and align legal judgment with moral intention rather than technical evasion. Under his administration, contemporaries noted that Chancery became so efficient it began drawing cases away from the common law courts, an innovation that earned Bacon both admiration and resentment. This pattern, too, is archetypal. In Raglan's formulation, the lawgiver frequently provokes opposition precisely because he restrains entrenched interests and unsettles habitual disorder. Like Solon or Lycurgus, Bacon's authority derived not from force or charisma, but from wisdom applied to chaos, and his reforms were aimed less at innovation than at stabilization.

The heroic lawgiver, in Lord Raglan's sense, is not a technician of custom but a prescriber of foundations. He appears before crisis, not after it; his task is to articulate in advance the laws by which a society might avoid its own destruction. It is therefore a category error to judge such figures by the standards of professional administration or legal conformity. Francis Bacon repeatedly distinguished the wisdom of the law-maker from that of the lawyer, and understood his own vocation in the former, not the latter, sense. That this self-conception was neither idiosyncratic nor retrospectively imposed is confirmed by the historical record. As Hugh Trevor-Roper observed, the reforms later demanded with violence in the 1640s had already been set out, lucidly and loyally, a generation earlier—not by Coke, but by Bacon:

> All the reforms of the law which would be loudly and angrily demanded by a rebellious people in the 1640s had been lucidly and loyally demanded, a generation before, not by Coke, but by Bacon. It was the same in education. Bacon, the greatest advocate of lay reason and lay religion, would have reformed the universities, dethroned Aristotle, introduced natural science; he would have stopped the growth of grammar schools and built up elementary education; he would have decentralised charitable foundations, whether schools or hospitals, for 'I hold some number of hospitals with competent endowments will do far more good than one hospital of exorbitant greatness'; he would have decentralised religion, planting and watering it in the forgotten 'corners of the realm'; and he would have decentralised industry, trade, wealth, for 'money is like muck, not good except it be spread.' When we read this evidence—evidence which is obvious, inescapable, constant throughout his writings—we can easily agree with the greatest of English seventeenth century historians, S.R. Gardiner, that if only Bacon's programme had been carried out, England would have escaped the Great Rebellion.[9]

When Bacon later remarked, in Latin, that he had foreseen civil disorder "on account of morals" (*propter mores*), he was not gesturing toward popular vice, doctrinal extremism, or abstract constitutional defect, but toward the visible corruption of governance at its center—a diagnosis already articulated, with remarkable consistency, in his essays "Of Counsel," "Of Nobility," and "Of Faction." In "Of Counsel," Bacon insists that stable rule depends upon structured, plural, and publicly intelligible deliberation; where access to the prince is monopolized by favourites, counsel degenerates into private confidence, and authority, however lawful in form, loses its moral visibility. "Of Nobility" develops the political consequences of this failure, locating the stability of monarchy in a respected and participatory aristocracy, and warning that the systematic bypassing of the nobility through personal intimacy, rather than public function, breeds resentment that is structural rather than personal. "Of Faction" completes the analysis, showing how such resentments, once normalized, detach loyalty from institutions and reattach it to persons, transforming private dependence into public division. Read together, these essays disclose a single causal chain: corrupted counsel produces alienation; alienation produces faction; faction, left uncorrected, produces civil conflict. Bacon's claim to have foreseen rebellion "on account of morals" is therefore neither prophetic nor rhetorical, but administrative and ethical in the strictest sense;

[9] Trevor-Roper, Hugh. *Religion, the Reformation and Social Change*, 1967, pp244-5

a judgment that the manners of power, long before the language of revolution emerged, had already made order untenable.

Ex malis moribus bonæ leges. 34

To the moſt iudicious, and learned, Sir FRANCIS BACON, Knight.

From *Minerva Brittana* (1612); the verses under Bacon read:

The Viper here, that stung the shepherd swain,
(While careless of himself asleep he lay,)
With Hyssop caught, is cut by him in twain,
Her fat might take, the poison quite away,
And heal his wound, that wonder tis to see,
Such sovereign help, should in a Serpent be.

By this same Leach, is meant the virtuous King,
Who can with cunning, out of manners ill,
Make wholesome laws, and take away the sting,
Wherewith foul vice doth grieve the virtuous still:
Or can prevent, by quick and wise foresight,
Infection ere it gathers further might.

This emblem from *Minerva Britanna* places Bacon squarely within what Lord Raglan identifies as the lawgiving phase of the heroic archetype. As Raglan observes, the hero of tradition is rarely remembered for administrative detail or political maneuvering; instead, "the only memorial of his reign... is the traditional code of laws which is often attributed to him," even though such laws are in fact the product of long historical evolution rather than individual invention. The heroic lawgiver does not create order *ex nihilo*. Rather, he intervenes at a moment of crisis, codifies, purifies, and restrains forces already at work, and then recedes. Peacham's shepherd, working not with sword or crown, but with measured incision, embodies precisely this function.

In the emblematic economy of *Minerva Britanna*, Bacon is figured as a modern analogue to the archaic lawgivers, above all Solon. The parallel is structural rather than rhetorical. Solon and Lycurgus impose foundational laws, refuse tyranny, withdraw or disappear, and leave their statutes to stand or fail without the prop of their continuing authority. They are remembered not for acts of rule but for acts of ordering. Bacon's self-placement within this lineage is explicit and sustained. In *A Proposition for Compiling and Amendment of Our Laws*, presented to James I around 1616, Bacon invokes the ancient legislators to underscore a principle central to his own project: laws must outlive their author, and their authority must not depend upon personal rule or reputation..."the laws of Lycurgus, Solon, Minos... are not the worse because grammar scholars speak of them."

That principle governs the substance of Bacon's reform proposals. Far from advocating radical innovation, Bacon calls for a systematic restoration and reordering of English law, what he describes as "pruyning and grafting," not uprooting. He diagnoses the causes of legal uncertainty with a jurist's precision: an excessive accumulation of statutes, ambiguity of language, disorderly exposition, and contradictions among judgments.

Against this chaos he proposes clarity, certainty, and accessibility. Inspired by the model of Justinian, Bacon urges the creation of a concise legal "digest": obsolete statutes removed, contradictions resolved, principles set out briefly but intelligibly, so that the law might be understood by ordinary subjects and not only by professional interpreters. In later documents—*An Offer to the King of a Digest to be Made of the Laws of England* (1621) and *The Elements of the Common Laws of England*—the same aim recurs: brevity without obscurity, clarity without prolixity, and institutions strong enough to function without continual personal intervention.

THE

ELEMENTS

OF THE COM-
MON LAWES OF
ENGLAND,

Branched into a double Tract:

THE ONE
Contayning a Collection of ſome princi-
pall Rules and Maximes of the Common
Law, with their Latitude and Extent.

Explicated for the more facile Introduction of ſuch as are ſtudiouſly addicted to that noble Profeſsion.

THE OTHER
The Vſe of the Common Law, for preſeruation
of our Perſons, Goods, and good Names.

According to the Lawes and Cuſtomes of this Land.

By the late Sir *Francis Bacon* Knight, Lo: Verulam
and Viſcount S. Alban.

Videre Vtilitas.

LONDON,
Printed by the Aſsignes of *I. More* Eſq. 1630.

Bacon's *Elements of the Common Laws of England* (1630)

Yet Bacon's most radical act of lawgiving occurs not in Westminster Hall but in the realm of knowledge itself, where the heroic pattern reaches its fullest expression. With the *Novum Organum,* Bacon does not merely propose a new scientific method; he promulgates a new constitution for the intellect. The work is saturated with legal metaphors that are neither casual nor decorative. Inquiry becomes a tribunal; nature is interrogated rather than speculated upon; conclusions are judgments rendered only after lawful procedure. The doctrine of the Idols functions as a sweeping repeal of corrupt statutes and illegitimate precedents that have governed human understanding: custom, authority, language, and theatrical philosophy standing in for entrenched but unlawful powers. True induction, by contrast, operates as due process: slow, cumulative, adversarial, and resistant to premature verdicts. Knowledge is no longer seized by brilliance or inspiration, but earned through obedience to law.

In Joseph Campbell's terms, this moment corresponds to the hero's "boon," the gift brought back from the ordeal for the benefit of the community. Yet Bacon's boon is distinctive. It is not a weapon, a secret, or a revelation, but a set of binding rules governing how truth itself may be sought. In this sense Bacon acts as a lawgiver in the strict Raglanian sense: he does not merely overthrow the old sovereign, Aristotle and the scholastic regime, but replaces it with binding norms meant to outlast his own authority. Forms function like legal definitions; tables of exclusion resemble evidentiary rules; method itself becomes statute. This juridical structure is not incidental. Bacon had spent a lifetime observing how judgment fails, how power corrupts procedure, and how systems harden into tyranny unless periodically re-founded. *Novum Organum* is thus best read as an act of epistemic legislation, binding the intellect under law so that future freedom might be possible.

That Bacon understood himself in this heroic lawgiving role is confirmed by contemporary reception and by his own words. In *Minerva Britanna* he is explicitly aligned with Solon, the archetypal lawgiver whose authority rested on wisdom rather than force. A generation later, the first historian of the Royal Society would elevate the analogy further, casting Bacon as a scientific Moses: one who leads others to the borders of the promised land of knowledge, though he does not himself enter it. This, too, follows the heroic pattern precisely: the lawgiver prepares the way, establishes the covenant, and withdraws, leaving the future to those bound by the laws he has given.

Most revealing of all is Bacon's insistence upon moral limits to knowledge itself. In *Valerius Terminus* he pauses before opening the "fountain" of new learning in order to erect "a strong and sound head or bank to rule and guide the course of the waters." Knowledge, he insists, must be "limited by religion, and referred to use and action," and even the smallest portion granted to humanity under God's "large charter" must remain subject to the benefit of society. This is not the voice of a Promethean rebel without restraint, but of a heroic legislator conscious that new power, intellectual as well as political, requires law lest it become destructive.

Taken together, Bacon's legal reforms, his epistemic legislation, and his self-conscious alignment with classical and biblical lawgivers complete a coherent heroic pattern. Like the shepherd in Peacham's emblem, Bacon confronts the serpent not to deny its danger, but to remove its sting; not to abolish human frailty, but to govern it wisely. In Raglan's terms, he leaves behind not monuments or conquests, but laws. In Campbell's, he brings back not fire alone, but the discipline to use it. Bacon's claim to heroic lawgiving thus rests not on symbolism alone, but on a sustained attempt to reorder both justice and knowledge at their foundations, binding even his own intellect under laws he hoped would endure beyond him.

Bacon, Shakespeare, and the law

The earliest biography of Bacon, written by his amanuensis and chaplain William Rawley, states that while Bacon became expert in the law, he preferred other studies in his youth:

> Being returned from travel, he applied himself to the study of the common law, which he took upon him to be his profession; in which he obtained great excellency, though he made that (as himself said) but as an accessory, and not his principal study. He wrote several tractates upon that subject: wherein, though some great masters of the law did out-go him in bulk, and particularities of cases, yet in the science of the grounds and mysteries of the law he was exceeded by none.

A letter from Bacon to his uncle Lord Burghley (William Cecil) pleads for a position that will enable him to study other things than law:

> Although it must be confessed that the request is rare and unaccustomed, yet if it be observed how few there be which fall in with the study of the common laws, either being well left or friended, or at their own free election, or forsaking likely success in other studies of more delight and no less preferment, or setting hand thereunto early without waste of years.

Poetaster (1601), a play written by Bacon's friend Ben Jonson, portrays a character named Ovid Junior, a law student who spends his time reading poetry instead of studying law, upsetting his father:

> *Ovid Sr.* Are these the fruits of all my travail and expenses? Is this the scope and aim of thy studies? Are these the hopeful courses wherewith I have so long flattered my expectation from thee? Verses? Poetry? Ovid, whom I thought to see the pleader [lawyer], become Ovid the play-maker?
>
> *Ovid Jr.* No, sir.
>
> *Ovid Sr.* Yes, sir; I hear of a tragedy of yours coming forth for the common players there, call'd *Medea* . . . What? shall I have my son a stager now?

Ovid Junior tells his father *"I am not known upon the open stage: nor do I traffic in their theatres."* Medea is of course known for killing her own sons; this seems to be a pointed reference to Elizabeth.

> The lawyer, when he reads attentively the works of William Shakespeare, may not be more surprised by the poet's correct use of law terms, and intimate acquaintance with legal customs and tenures and the *lex scripta,* than by his extensive and profound knowledge of the maxims of the English law.
>
> —William Lowes Rushton, *Shakespeare's Legal Maxims*

Many writers have claimed that Shakespeare's extensive and exact legal knowledge strongly suggests he was a lawyer; as Mark Twain put it in his book *Is Shakespeare Dead?* (1909):

> Shakespeare couldn't have written Shakespeare's works, for the reason that the man who wrote them was limitlessly familiar with the laws, and the law-courts, and law-proceedings, and lawyer-talk, and lawyer-ways . . . a man can't handle glibly and easily and comfortably and successfully the argot of a trade at which he has not personally served. He will make mistakes; he will not, and cannot, get the trade-phrasings precisely and exactly right; and the moment he departs, by even a shade, from a common trade-form, the reader who has served that trade will know the writer hasn't.

A striking feature of Shakespeare's engagement with law, most vividly in *The Merchant of Venice* and *Measure for Measure,* is that it operates not merely at the level of legal vocabulary or courtroom spectacle, but at the level of professional legal reasoning. Law in these plays functions as an internally coherent system whose logic, when pressed to its limits, produces both justice and catastrophe. This is precisely the level at which Francis Bacon theorized law. In his essay "Of Judicature," Bacon repeatedly warns against the danger that rigid legality poses when severed from judgment, cautioning judges "that that which was meant for terror be not turned into rigor," especially when "penal laws... have been sleepers of long." The same concern appears even more pointedly in Bacon's private counsel to Elizabeth, where he observes that "the cessation and abstinence to execute these unnecessary laws do mortify the execution of such as are wholesome." Law neglected, Bacon insists, does not simply lapse; it corrodes the authority of law itself.

This exact anxiety is dramatized in *Measure for Measure*. Vienna's ruler confesses that "We have strict statutes and most biting laws... / Which for this fourteen years we have let slip," only to discover that sudden enforcement produces injustice rather than order. Shakespeare gives this problem a memorable image: "In time the rod becomes more mocked than feared." The metaphor aligns perfectly with Bacon's diagnosis. Law unenforced becomes contemptible; law enforced without discretion becomes cruel. The danger lies not in law itself, but in the oscillation between neglect and severity, a pattern Bacon feared would "mortify" even those statutes intended for the public good.

The affinity deepens in both writers' treatment of equity and mercy. Bacon insists that no legal system can survive without equity, invoking the maxim *summum jus, summa injuria* (extreme law is extreme injustice) and reminding judges that their office is *jus dicere*, not *jus dare*: "to interpret law, and not to make law." Mercy, for Bacon, is not sentimental indulgence but a judicial virtue exercised "as far as the law permitteth," with "a merciful eye upon the person," even while maintaining the force of legal example. This logic governs the trial scene in *The Merchant of Venice*. Portia does not defeat Shylock by overriding the law with compassion; she defeats him by embracing the bond absolutely and then interpreting it more strictly than its author anticipated. Mercy is praised: "The quality of mercy is not strained," but the verdict itself turns entirely on construction, statute, and jurisdiction. Law is not suspended; it is turned back upon itself.

Taken together, these parallels do not prove common authorship. But they do establish a shared legal imagination, one deeply shaped by Inns of Court culture, in which law is simultaneously a civilizing necessity and a perilous abstraction. In Bacon's essays and memoranda, and in Shakespeare's plays, the same fear recurs: that law, when either forgotten or fetishized, ceases to protect and instead becomes a machinery of forfeiture, turning words into bonds and justice into sacrifice. The question, then, is not whether Shakespeare "knew some law," but why the plays think like law under pressure, and why that thinking so closely resembles the legal philosophy articulated by Francis Bacon. What remains is not a verdict, but a record. And records, in law as in history, have a way of speaking for themselves.

> The abstrusest elements of the common law are impressed into a disciplined service with every evidence of the right and knowledge of commanding. Over and over again, where such knowledge is unexampled in writers unlearned in the law, Shakespeare appears in

perfect possession of it. In the law of real property, its rules of tenure and descents, its entails, its fines and recoveries, and their vouchers and double vouchers; in the procedure of the courts, the methods of bringing suits and of arrests, the nature of actions, the rules of pleading, the law of escapes, and of contempt of court; in the principles of evidence, both technical and philosophical; in the distinction between the temporal and spiritual tribunals; in the law of attainder and forfeiture; in the requisites of a valid marriage; in the presumption of legitimacy; in the learning of the law of prerogative; in the inalienable character of the crown, this mastership appears with surprising authority.

—Cushman K. Davis, *The Law in Shakespeare*

Queen Elizabeth in Parliament

8. The hero falls

The publication of Bacon's *Novum Organum* in 1620 marks one of the most radical moments in the intellectual history of early modern Europe. Bacon does not merely revise Aristotle; he annuls him. He does not offer an improvement to inherited learning; he announces a new beginning. The very title proclaims replacement: a new instrument of thought, a new discipline, a new covenant between the human mind and nature. The work is legislative in tone and ambition. It lays down binding procedures for inquiry and rejects the authority of tradition, custom, and premature abstraction. Bacon writes not as a commentator but as a lawgiver for knowledge.

What follows immediately after is therefore impossible to ignore. In the spring of 1621, scarcely a year after the appearance of *Novum Organum*, Bacon, then Lord Chancellor, at the apex of political authority, was charged with twenty-three counts of bribery and corruption. He was impeached, briefly imprisoned in the Tower of London, fined, and formally barred for life from office, Parliament, and proximity to the Court. From an administrative standpoint, this sequence can be explained by factional rivalry, procedural opportunism, and the volatile mechanics of Jacobean politics. From an archetypal standpoint, however, the pattern is exact.

Bacon's fall is especially resonant because it is not the punishment of a heretic or rebel. He is not condemned for sedition, heterodoxy, or treason. He is condemned for corruption, precisely the charge that undermines the moral authority of a judge and lawgiver. In mythic terms, this is not the execution of an enemy but the ritual disqualification of a priest-king. The penalties imposed were severe on paper: a fine of £40,000, imprisonment at the king's pleasure, permanent incapacity to hold office, exclusion from Parliament, and banishment from the verge of the Court. Although many of these penalties were later mitigated (Bacon was released after two days, pardoned by James, and the fine quietly neutralized), the symbolic damage was irreversible. His public authority was destroyed.

What matters archetypally is not the precise measure of Bacon's legal guilt, which historians continue to debate. Early modern judicial practice relied heavily on customary gift-giving; administrative dysfunction was endemic; and Bacon's prosecution was orchestrated by his lifelong rival, Sir Edward Coke, who revived an impeachment procedure that had not been used in over a century. What matters is that the figure who had just proclaimed a new intellectual order was expelled from the existing one.

This pattern is ancient. Moses gives the law but does not enter the promised land. Prometheus gives fire but does not enjoy it. Oedipus uncovers the truth but cannot rule with it. Solon gives Athens its laws and then departs. The reformer survives, but only as exile. Authority is severed from office so that law may persist without the man.

The archetypal resonance deepens when Bacon's own words are considered. Writing to the king during the crisis, he declared: "I wish that as I am the first, so I may be the last of sacrifices in your time." This is not the language of a bureaucrat defending himself against procedural error; Bacon does not protest innocence, he accepts immolation in the hope that the system itself might be purified. The phrasing casts him not as a defendant but as a victim offered to stabilize a larger order.

From this perspective, the fall of 1621 does not negate *Novum Organum*; it completes its heroic logic. The work must be separated from the man, the method must outlive the reputation. Authority must be destroyed so that influence can become universal. Had Bacon fallen before 1620, *Novum Organum* would read as the manifesto of a disgruntled outsider. Had he fallen long after, it would read as the product of a successful career. Falling one year later transforms the book into something closer to a testament; not consciously a last will, but structurally one all the same.

The legal circumstances themselves reinforce the pattern. Despite his reputation, Bacon was not notably wealthy. At the height of his practice, he earned four to five thousand guineas a year, modest compared to contemporaries, some of whom earned ten times as much. He lived beyond his means, remained in debt, and showed little evidence of a personality driven by financial accumulation. More importantly, Bacon had been actively working to reform the very fee-based system under which judges were paid. Through speeches and writings, he argued for replacing uncertain, irregular payments with fixed state salaries, warning that an unpaid bench might become a dangerous power in a highly litigious society.

This reform agenda directly threatened entrenched interests, particularly those of Coke, who prospered under the existing system. As Lord Chancellor, Bacon worked at extraordinary pace, clearing a massive backlog of cases within months despite chronic illness. Over four years he delivered judgment in approximately 8,000 cases. His efficiency drew litigants away from the common law courts into Chancery, depriving other judges and officials of lucrative fees. If so, Bacon's reforms were not merely theoretical; they had immediate financial consequences for his rivals.

Several writers have therefore defended Bacon's record as Chancellor, arguing that his prosecution cannot be understood apart from rivalry, resentment, and institutional resistance to reform. But it is not necessary to resolve these questions in order to grasp the archetypal shape of the event. Bacon was "driven from throne and city." The lawgiver was expelled from the polity he sought to reform.

From this point on, Bacon's life takes on a new clarity. He continues to write, but no longer for contemporaries. His authority becomes posthumous, his audience posterity. He is no longer a ruler within institutions but a founder whose work requires his own removal. This figure is among the most stable in heroic tradition: the reformer of knowledge rather than armies, punished by social death rather than execution, vindicated only in the future.

9. The hero dies a mysterious death

Francis Bacon's death has long been framed as the final emblem of his devotion to experimental inquiry. According to the familiar account, he caught pneumonia in early April while conducting an experiment on the preservation of flesh, stuffing a chicken carcass with snow. The story is irresistible in its symbolism: the philosopher of experiment dying in the act of experiment, a martyr to the new science.

Yet the very neatness of the anecdote invites scrutiny. Early April is an unlikely time for sufficient snow to be readily available, and Bacon had already written theoretically about cold as a preservative, raising questions about whether the episode describes a genuine experiment or a retrospective construction. Lisa Jardine and Alan Stewart, in *Hostage to Fortune: The Troubled Life of Francis Bacon*, caution against taking such deathbed stories at face value. As they observe:

> Accounts of the circumstances surrounding a prominent death in early modern England need to be taken with more than a pinch of salt. Just like the anecdote of Sir Nicholas Bacon dispensing his last *bon mot* on the barber who thoughtfully left open a window for fresh air (that contained the 'cold' that killed him), this account of Bacon's end is carefully constructed.

They suggest that the experimental anecdote may have served to mask a more prosaic or embarrassing cause of death, possibly an overdose of opiate. Whether or not this speculation is correct, the important point is that Bacon's death was narrativized almost immediately, shaped into a form that aligned with his public persona and intellectual legacy.

In this respect, Bacon's end closely resembles that of Pliny the Elder, who died in 79 CE during the eruption of Vesuvius. Pliny, author of the *Natural History* and an indefatigable compiler of empirical knowledge, sailed toward the eruption both to observe the phenomenon and to aid those in danger. He was overcome by volcanic fumes and died on the shore near Stabiae. Like Bacon, Pliny became an emblematic figure: the scholar who dies not in retreat but in pursuit of understanding. His death, recorded by his nephew Pliny the Younger, functions as a founding legend for empirical inquiry, blurring the line between historical event and exemplary story.

The parallel is instructive. In both cases, the death serves less as a medical report than as a symbolic closure. Pliny's death consecrates Roman natural history; Bacon's death consecrates the experimental method. Whether or not the precise circumstances unfolded as later accounts describe, the form of the story matters. The hero of knowledge does not simply die; he dies *in character*. This is precisely the pattern Raglan identifies: the hero's death is often obscure, contested, or ritualized, and its meaning lies not in forensic certainty but in narrative function.

Seen in this light, Bacon's "martyrdom to science" need not be literally true to be structurally significant. Like Pliny before him, Bacon is positioned at the threshold between old and new regimes of knowledge. The ambiguity surrounding his death mirrors the ambiguity of the transition he represents. What survives is not the body, but the story: the image of a lawgiver of knowledge whose life ends in proximity to the very forces he sought to understand.

Bacon's will bequeaths his papers as follows:

> Take care that of all my writings, both of English and of Latin, there may be books fair bound, and placed in the King's library, and in the library of the University of Cambridge, and in the library of Trinity College, where myself was bred, and in the library of Bene's College [Corpus Christi, Cambridge], where my father was bred, and in the library of the University of Oxford, and in the library of my Lord of Canterbury, and in the library of Eton. Also, I desire my executors, especially my brother Constable, and also Mr. Bosvile, presently after my decease, to take into their hands all my papers whatsoever, which are either in cabinets, boxes, or presses, and tell them to seal up until they may at their leisure peruse them.

Of course, the Stratford will makes no mention of any books or papers.

Bacon's death fits another criterion of the Raglan pattern, specifically the motif of the hero who dies at the top of a hill. Francis Bacon died at Highgate, then a rural elevation just north of London, while lodging at the house of Earl of Arundel. Highgate, rising above the city, had long been associated with health, air, and retreat; it was a liminal zone between court and countryside, between the civic world Bacon had lost and the contemplative world he increasingly inhabited in his final years. In mythic terms, the ascent to a height before death is rarely accidental: hills and mountains mark thresholds, places where the human approaches the divine, where vision is clarified even as the body fails.

The Arundel residence associated with Bacon's final days would later become one of the early meeting places of the Royal Society after the Great Fire of London displaced institutions across the city. The symbolism is difficult to ignore: Bacon dies on a height, removed from power, yet the institutional embodiment of his program takes root in the orbit of the very household that sheltered his final days. Myth functions here as a bridge between the divine and the human, but also between individual life and collective destiny. Elevated places recur across traditions not merely as scenic backdrops but as sites of disclosure and transition: Sinai, Olympus, Calvary, the Acropolis. Bacon's death at Highgate does not prove anything, nor need it be overburdened, but it quietly fulfills the pattern. The lawgiver of a new epistemology delivers his code, falls from worldly authority, and exits history from a height, leaving behind not a throne or a dynasty, but an institution and a method that would reshape the world below

Raglan's final points turn from the drama of death to the paradoxical afterlife of the hero, and here again Francis Bacon fits the pattern with striking fidelity. Bacon was born at York House in London, but his life and memory are decisively anchored in St Albans, where he styled himself Viscount St Alban and where his monument stands in St Michael's Church. The monument, erected by his devoted secretary William Rawley, presents Bacon seated in contemplative posture, very much alive in thought, though dead in body: an image closer to a lawgiver or sage than to a dynastic nobleman. It is a memorial to intellect rather than lineage, already anticipating Raglan's next observation.

Bacon left no legitimate children to succeed him. His title became extinct at his death; his line, in the ordinary biological sense, ended with him. Yet this absence sharpens the archetypal reading. The hero of myth often founds not a family, but a tradition; not a bloodline, but a discipline. Bacon's "children" are his books, his methods, his institutions, most notably the future experimental community that would gather under the banner of the Royal Society. In Raglan's terms, succession fails at the level of flesh, only to reappear at the level of culture.

The final paradox concerns burial and the multiplication of tombs. Bacon is said to be interred at St Michael's, but doubt remains as to whether his remains are actually there, and whether death was faked.

The title page of *Minerva Britanna* (1612) features a writing hand emerging from behind the curtain of a theatre proscenium. Entwined among laurel leaves, the legend translated reads "One lives in one's genius, other things depart in death." The hand has just written *mente videbor*, "By the mind I shall be seen."

10. Francis Bacon and Shakespeare

> "Lord Bacon was the greatest genius that England, or perhaps any country, ever produced."
>
> —Alexander Pope

My Shakespeare skepticism began with reading *Richard III* and then *Richard II* in close succession. *Richard III* is rhetorically assured; *Richard II*, by contrast, is hesitant in places, even clumsy. That difference makes sense once the political stakes of *Richard III* are acknowledged. The Tudor dynasty derived its legitimacy from Richard's overthrow; to secure that legitimacy, Richard could not merely be defeated, he had to be damned. The play does precisely this work. Richard is rendered monstrous from birth, stripped of ambiguity and positioned as an affront to divine order itself. His fall is not simply historical but providential. The Tudors inherit not a contested throne, but a moral restoration. In this sense *Richard III* is not merely drama, but a major act of statecraft: a theatrical extension of Tudor historiography, legitimating regime change by transfiguring politics into theology.

I didn't know all that, of course, but years earlier a high-school teacher had remarked that *Romeo and Juliet* was written by the Earl of Oxford. At the time it meant nothing. Only later did it reappear, not as a claim demanding belief, but as an invitation to investigate. Once the assumption of a single, seamless authorial voice loosens, attention shifts from personalities to pressures: who wrote these plays, under what conditions, and at what risk.

From there the path to Bacon was rapid. Bacon matters because he understood, more clearly than anyone of his age, the intimate relationship between knowledge, power, legitimacy, and controlled disclosure. The Shakespearean histories and Bacon's philosophical project inhabit the same symbolic terrain: both are concerned with lawgiving, succession, rebellion, sovereignty, and the dangerous necessity of indirection in a political culture where speech itself could be fatal.

In this light, Shakespeare begins to resemble something like England's moon landing: an achievement so vast, so central to national identity, that it requires an official story to hold it in place. Everyone knows the story; few examine its engineering. This book does not ask the reader to abandon the canon, only to read it with historical nerves intact; to notice where drama serves power, where it strains against it, and where the fractures of state necessity leave their marks in art.

> His meaning had not escaped me. If you analyzed it, it was the old Bacon and Shakespeare gag. Bacon, as you no doubt remember, wrote Shakespeare's stuff for him and then, possibly because he owed the latter money or it may be from sheer good nature, allowed him to take the credit for it. I mentioned this to Jeeves, and he said that perhaps an even closer parallel was that of Cyrano de Bergerac.
>
> P.G. Wodehouse, *Joy in the Morning*

> "You must have heard of Shakespeare. He's well known. Fellow who used to write plays. Only Aurelia's aunt says he didn't. She maintains that a bloke called Bacon wrote them for him." "Dashed decent of him,"

said Archibald, approvingly. "Of course, he may have owed Shakespeare money." "There's that, of course."

P.G. Wodehouse, "The Reverent Wooing of Archibald"

Contemporaries of Bacon eulogized him as "a muse more choice than the nine muses" who "showered the age with frequent volumes" and "filled the world with works"; "the very nerve of genius, the marrow of persuasion, the golden stream of eloquence, the precious gem of hidden literature."[10] Ben Jonson's *Timber*, published posthumously in 1641, states that Bacon "performed that in our tongue which may be compared, or preferred, either to insolent Greece or haughty Rome . . . he may be named and stand as the mark and acme of our language." This is somewhat puzzling, as Bacon published just three books in English during his life, the *Essays* (in successively expanded editions, 1597, 1612, 1625), *The Advancement of Learning* (1605), and *The History of the Reign of King Henry VII* (1622). Even more strangely, Jonson (who was sparing in his praise of other writers) had already bestowed the same encomium upon Shakespeare in the 1623 First Folio:

> Leave thee alone for the comparison
> Of all that insolent Greece or haughty Rome
> Sent forth, or since did from their ashes come.

An early claim that Bacon wrote anonymously occurred in 1599.[11] After English authorities published an account of a plot to poison Queen Elizabeth's saddle, an anonymous writer ascribed the state's "smooth penned pamphlet" to "*M. Smokey-swynes flesh,* at the instance of Sir R.C." (i.e., Francis Bacon writing at the behest of Secretary of State Robert Cecil). This is again rather strange, as Bacon had only published a small book of ten essays and some religious meditations in his own name, very little to inform attribution of other tracts. Thomas Tenison, who crowned two monarchs as Archbishop of Canterbury, edited a volume of Bacon's previously unpublished materials in 1679, in which he wrote "those who have true skill in the works of the Lord Verulam, like great masters in painting, can tell by the design, the strength, the way of colouring, whether

[10] Rawley, William. *Memoriæ honoratissimi domini Francisci, Baronis de Verulamio, Vice-Comitis Sancti Albani sacrum.* London: John Haviland, 1626. (Translated from Latin)
[11] Stewart, Alan. "Rethinking Authorship Through Collaboration" in *Renaissance Transformations: The Making of Writing in Renaissance England.* Margaret Healy and Thomas Healy (eds.). Edinburgh: Edinburgh University Press, 2009.

he was the author of this or the other piece, though his name be not to it."[12] A period of twenty years separates the time when Bacon left Cambridge in 1576 to when his name first appeared in print, with the publication of his *Essays* in 1597; yet, according to Brian Vickers, editor of the Oxford *Major Works*, Bacon had a substantial body of anonymous work and was reluctant to have his name in print:

> Although Bacon had a wide and diverse literary output by 1597 (enough to fill several hundred pages of Spedding's edition of his *Occasional Works*), none of it had been publicly acknowledged as his composition. Indeed, it was only due to an impending plagiarization that his name finally appeared in print that year, to forestall the unauthorized publication of his *Essays* . . . Had Serger not attempted his unauthorized edition, Bacon's first appearance in print might have been as the author of *The Advancement of Learning* (1605), that bold attempt to persuade King James to initiate a total reformation of study and research in his new kingdom.[13]

Bacon's writings may have been collaborative to a considerable extent. The article "Who Wrote Bacon? Assessing the Respective Roles of Francis Bacon and His Secretaries in the Production of His English Works" applies modern stylometric (wordprint) analysis to a long-standing puzzle in Bacon scholarship: how Bacon could have produced such a vast, stylistically inconsistent body of work while maintaining an exceptionally demanding political career. Using two independent statistical methods, the authors compare Bacon's handwritten letters (autographs) with his published English works and with texts by contemporaries, including Thomas Hobbes. The central finding is stark: Bacon's autographic writings display a consistent, identifiable wordprint, but most of the works published under his name do not match it.

The authors argue that the most economical explanation for this discrepancy is Bacon's extensive reliance on secretaries and collaborators. While Bacon clearly possessed a stable personal writing style, detectable in his handwritten letters, that style is largely absent from his published corpus. Statistical comparisons show that over half of Bacon's published English works differ so strongly from his autograph wordprint that they are unlikely to have been written solely by him. Only one major published work—the Apology Concerning the Earl of Essex—consistently matches

---

[12] Tenison, Thomas. *Baconiana, or, Certain genuine remains of Sr. Francis Bacon, Baron of Verulam, and Viscount of St. Albans in arguments civil and moral, natural, medical, theological, and bibliographical now for the first time faithfully published*. London, 1679.

[13] Vickers, Brian. "The Authenticity of Bacon's Earliest Writings." *Studies in Philology* 94, no. 2 (1997): 248-96.

Bacon's autograph style, suggesting it was composed directly by Bacon during a moment of personal and political crisis.

The study further explores the possibility that Thomas Hobbes, who served intermittently as Bacon's secretary, may have contributed substantially to certain texts, particularly portions of *New Atlantis* and some historical writings. While the authors stress that stylometric evidence cannot conclusively prove authorship in cases of collaboration, several Bacon texts cluster closely with Hobbes's known wordprint. This overlap is strongest during periods when Hobbes is historically documented as having access to Bacon, lending circumstantial support to the hypothesis of significant Hobbesian involvement, especially in early and middle sections of *New Atlantis*.

Importantly, the authors reject sensational conclusions. They do not argue that Bacon lacked originality or that his philosophical ideas were authored by others. Instead, they propose a model of distributed authorship, common in early modern political and intellectual life, in which Bacon functioned as an originator, director, and editor of ideas that were often drafted, phrased, or expanded by secretaries. Historical evidence, ranging from household records to John Aubrey's accounts of Bacon dictating ideas during walks, strongly supports this picture of collaborative production.

The article concludes that recognizing Bacon's reliance on secretarial assistance resolves several long-standing anomalies: contradictions in style, shifts in tone, and the sheer volume of output. Rather than undermining Bacon's significance, this perspective reframes him as an intellectual architect whose ideas were realized through a collective process. The authors suggest that future scholarship should focus less on defending single authorship and more on reconstructing the institutional and collaborative conditions under which Bacon's thought was produced.

Alan Stewart, co-chair of the *Oxford Francis Bacon* and co-author of the biography *Hostage to Fortune*, describes a scenario whereby

> Surviving autograph drafts by Bacon—that is, tracts written in his own handwriting—are scant indeed, and most of them are notes, occurring in commonplace books and notebooks . . . the norm is a draft in the hand of one of Bacon's secretaries or amanuenses, with Bacon's autograph comments, often quite extensive emendations . . . That these writings range

> in date from 1603 to 1621 suggests that Bacon's preferred form of writing may always have been collaborative.[14]

The Workes of Benjamin Jonson (1616) is seen as a forerunner of the Shakespeare First Folio, being the first time an English playwright's works were issued in a large, expensive folio edition. Evidently it was a collaborative effort; Jonson's statement that he wrote *Volpone* "without a co-adjutor, novice, journey-man or tutor" implies that

> The main author reigned above co-adjutors (helpers or assistant writers), novices (inexperienced or probationary writers) and journeymen (writers who were newly qualified, having finished their apprenticeships). The definitions of all three positions . . . imply that each served in subservient positions to more experienced masters, such as Jonson.[15]

Jonson, a close friend and occasional assistant of Bacon, wrote in *Timber, or Discoveries Made upon Men and Matter* (1641):

> He [Francis Bacon] hath filled up all numbers, and performed that in our tongue which may be compared or preferred either to insolent Greece or haughty Rome. In short, within his view, and about his times, were all the wits born that could honor a language or help study. Now things daily fall, wits grow downward, and eloquence grows backward; so that he may be named and stand as the mark and acme of our language.

Notice that Jonson, a playwright, identified Bacon, not Shakespeare, as the "mark and acme of our language."

> I remember once lunching with rare Ben Jonson at the Mermaid Tavern—this would be back in Queen Elizabeth's time, when I was beginning to be known in the theatrical world—and seeing a young man with a nobby forehead and about three inches of beard doing himself well at a neighboring table at the expense of Burbage the manager. "Ben," I asked my companion, "who is that youth?" He told me that the fellow was one Bacon, a new dramatist who had learned his technique by holding horses' heads in the Strand, and who, for some reason or other, wrote under the name of Shakespeare. "You must see his *Hamlet*," said Ben enthusiastically. "He read me the script last night. They start rehearsals at the Globe next week. It's a pippin. In the last act every blamed

[14] Stewart 2009.

[15] Grace Ippolo, *Dramatists and their Manuscripts in the Age of Shakespeare, Jonson, Middleton, and Heywood: Authorship, Authority and the Playhouse* (London: Routlede, 2006), p. 32.

character in the cast who isn't already dead jumps on everyone else's neck and slays him."

P.G. Wodehouse, "My Life as a Dramatic Critic"

The Story of the Learned Pig (1786) playfully alludes to Bacon's descent, plainly stating he was behind the Shakespeare works:

> My parents, indeed, were of low extraction; my mother sold fish about the streets of this metropolis, and my father was a water-carrier celebrated by Ben Jonson in his comedy of *Every Man in his Humour* . . . I soon after contracted a friendship with that great man and first of geniuses, the 'Immortal Shakespeare,' and am happy in now having it in my power to refuse the prevailing opinion of his having run his country for deer-stealing, which is as false as it is disgracing. The fact is, Sir, that he had contracted an intimacy with the wife of a country Justice near Stratford, from his having extolled her beauty in a common ballad; and was unfortunately, by his worship himself, detected in a very awkward situation with her. Shakespeare, to avoid the consequences of this discovery, thought it most prudent to decamp. This I had from his own mouth. With equal falsehood has he been father'd with many spurious dramatic pieces. *Hamlet, Othello, As You Like It,* the *Tempest,* and *Midsummer's Night Dream,* for five; of all which I confess myself to be the author.

While the *Learned Pig* does not specifically mention Bacon by name, the "water-carrier celebrated by Ben Jonson" is a character named Cob; when he appears onstage the following exchange takes place:

> *Cob.* I sir, I and my linage ha' kept a poor house, here, in our days.
> *Mat.* Thy linage, Monsieur Cob, what linage, what linage?
> *Cob.* Why Sir, an ancient linage, and a princely. Mine ance'try came from a King's belly, no worse Man
> . . .
> *Cob.* I Sir, with favour of your Worship's nose, Mr. Matthew, why not the ghost of a herring Cob, as well as the ghost of rasher-bacon?
> *Mat.* Roger Bacon, thou wouldst say?
> *Cob.* I say Rasher-Bacon. They were both broil'd o' the coals; and a man may smell broil'd meat, I hope? you are a scholar, upsolve me that, now…
> . . .
> *Mat.* Lie in a water-bearer's House! A Gentleman of his havings! Well, I'll

> tell him my mind.

Bacon was born in the sign of Aquarius, or the house of the water bearer; in this connection it might be worthwhile to cite the oldest representative of the heroic archetype, that of Sargon of Akkad, founder of Babylon:

> Sargon, the mighty king, King of Agade, am I. My mother was a vestal, my father I knew not, while my father's brother dwelt in the mountains. In my city Azuripani, which is situated on the bank of the Euphrates, my mother, the vestal, bore me. In a hidden place she brought me forth. She laid me in a vessel made of reeds, closed my door with pitch, and dropped me down into the river, which did not drown me. The river carried me to Akki, the water carrier. Akki the water carrier lifted me up in the kindness of his heart. Akki the water carrier raised me as his own son, Akki the water carrier made of me his gardener. In my work as a gardener I was beloved by Ishtar, I became the king, and for forty-five years I held kingly sway.

The Smedley books acquired by the Folger Shakespeare Library

William Smedley occupies an important, if now largely forgotten, place in the early twentieth-century afterlife of Francis Bacon. As editor of *Baconiana* for roughly a decade, Smedley functioned less as a polemicist than as a custodian of materials. He assembled an extensive private library of early modern books, philosophical, literary, theological, many bearing marginalia and corrections in Bacon's handwriting. After Smedley's death, a significant portion of this collection entered the Folger Shakespeare Library, where it remains underutilized. Among the most intriguing items is a copy of *The Anatomie of the Minde* (1576), alleged to contain corrections and annotations attributable to Bacon. Whether or not such attributions ultimately withstand paleographical scrutiny, the intellectual profile of the annotations—legal, moral, psychological—aligns closely enough with Bacon's known interests to warrant renewed, non-partisan examination using modern imaging and handwriting-comparison techniques. In *The Mystery of Francis Bacon* Smedley wrote:

> The following suggestion is put forward with all diffidence, but after long and careful investigation. Francis Bacon was the author of two books which were published, one before he left England, and the other shortly after. The first is a philosophical discourse entitled *The Anatomie of the Minde*. . . There is in existence

a copy of the book with the printer's and other errors corrected in Bacon's own handwriting.

Smedley did not identify the book he thought Bacon published soon after he left England, but I believe it's the book known as *Anti-Machiavel*. That book and *The Anatomie of the Minde* have extensive parallels with each other, documented in an appendix here, and appear to have been composed during Bacon's time at Cambridge.

A parallel case with Smedley's Bacon library is the collection assembled by Edwin Durning-Lawrence, whose books are now housed at the Senate House Library. Durning-Lawrence, a wealthy and indefatigable Bacon enthusiast, amassed one of the most substantial private Baconian libraries of the period. Among its highlights is a 1612 English *Don Quixote* containing marginal corrections and annotations that he likewise attributed to Bacon. As with the Smedley materials, the scholarly problem is not whether one must accept these attributions wholesale, but that the books themselves, objects bearing layers of early modern readerly intervention, have never been systematically digitized, collated, or subjected to high-resolution multispectral analysis. A focused project to digitize and cross-compare the Smedley and Durning-Lawrence holdings would allow scholars to move beyond anecdote and advocacy toward a genuinely empirical assessment: comparing hands across multiple volumes, mapping patterns of annotation, and situating them within the broader ecology of Elizabethan and Jacobean marginal practice. Even if Bacon's direct authorship were ultimately disproved in specific cases, the exercise would still recover a lost archive of elite early modern reading, one that speaks directly to the intellectual environment out of which Bacon's works emerged.

11. *Hamlet* and the heroic archetype

Hamlet is often approached as a revenge tragedy, a philosophical drama, or a case study in psychological paralysis. But it is also something deeper and older: a mythic narrative in tragic form. The play is saturated with archetypal structures that long predate Shakespeare—royal succession crisis, the slain father, the usurper on the throne, the summons to vengeance, the polluted kingdom, the sacrificial clearing of the ground for renewed order. Yet Shakespeare does not merely repeat the heroic pattern. He mutates it. In *Hamlet*, heroism is relocated from feats of strength to feats of conscience; from outward conquest to inward differentiation; from the confident performance of a role to the agonizing discovery that inherited roles no longer guarantee moral legitimacy.

This is why Hamlet is best understood as a failed king but a successful hero. He never reigns, founds no dynasty, and leaves no heir. And yet the play's action turns upon him as upon the indispensable instrument of purification. He exposes illegitimate power, forces hidden guilt into visibility, and absorbs into himself the disorder of the realm until it can no longer be contained. The kingdom is healed not by his accession but by his sacrifice. If the traditional hero's reward is the throne, Hamlet's reward is something more austere: the grim achievement of truth, purchased at the cost of life.

Modern archetypal theorists provide a language for this transformation. Otto Rank, writing in the early twentieth century, already sensed that *Hamlet* could be read as an archetypal drama in which the heroic demand—patricide, succession, the displacement of a father—has become psychologically self-aware and therefore intolerable. Lord Raglan's royal hero pattern aligns with the play's basic structure: royal birth, murdered king, usurpation, fatal reckoning, while also illuminating its decisive deviation: the hero does not become king. Joseph Campbell's monomyth clarifies the inversion of the hero's journey: the call comes (the Ghost), refusal and delay follow, trials accumulate, and the "boon" is finally delivered—yet the return is not triumphant, but terminal. And in the Jungian line, especially in Erich Neumann's account of ego separation from archaic authority, Hamlet becomes the emblem of a modern hero: one who must differentiate from the commanding Father not by blind obedience, but by ethical scrutiny, refusing to collapse into the old blood-logic even when commanded by the dead.

This heroic mutation is not merely psychological; it is historical. By Shakespeare's time, the symbolic grammar that once grounded medieval sovereignty had been destabilized. Authority could no longer rely unquestioned on sacred continuity and inherited role; it had to justify itself under conditions of uncertainty, surveillance, and moral doubt. *Hamlet* dramatizes what it is like to live inside that vacuum. The hero inherits the script of vengeance, but no longer trusts its premises. He is asked to act, but he cannot act cleanly; he is called to restore legitimacy, but legitimacy itself has become problematic. In that sense Hamlet is not an anti-hero, he is the heroic archetype after its foundations have cracked: the figure in whom the mythic demand persists, but conscience refuses to become its instrument without warrant.

Archetype is the play's hidden engine. The political plot is the surface; the hero-pattern is the deep grammar. Shakespeare takes traditional heroic materials (ghost, usurper, poisoned court, avenging son), and subjects them to an unprecedented inward pressure: reflection, hesitation, moral recoil, awareness of contamination. The result is a tragedy that does not abolish heroism, but transposes it. Hamlet becomes the hero of consciousness: a prince who does not conquer the world, but who discovers, at terrible cost, what it means to act justly when the world itself has become suspect.

Royal succession and the broken pattern (Raglan)

If Rank shows us why Hamlet cannot psychologically fulfill the heroic script, Lord Raglan helps explain why he cannot fulfill it politically. In *The Hero: A Study in Tradition, Myth, and Drama,* Raglan identifies a recurring biographical pattern shared by mythic heroes and sacred kings: noble birth, a crisis or threat surrounding the father, displacement or exile, the hero's eventual triumph, accession to rule, and a mysterious or violent death that often secures fertility or continuity for the realm. The pattern is not merely narrative; it is structural, linking heroism to kingship and kingship to cosmic order.

Measured against this template, *Hamlet* fits remarkably well—until the moment when it does not.

Hamlet is of royal birth. His father is king. The father is murdered under suspicious circumstances. A usurper occupies the throne. The hero's task is to restore legitimacy by destroying the illegitimate ruler. Up to this point, the play tracks the royal hero pattern almost mechanically. Shakespeare even provides parallel "control" figures, Laertes and Fortinbras, who move through similar father-avenging scripts with greater speed and less hesitation, as if to remind us what the traditional pattern looks like when it proceeds unchecked.

But Hamlet never completes the Raglan arc. He does not marry. He does not reign. He does not found a dynasty. The heroic function is fulfilled, but the heroic reward is withheld.

This deviation is not incidental; it is the play's deepest structural statement. In Raglan's schema, the hero's death is often meaningful precisely because it occurs after kingship has been secured or symbolically ratified. Hamlet's death, by contrast, occurs instead of kingship. Succession passes not to the hero but around him, to Fortinbras, a figure who has not undergone the play's moral ordeal. Hamlet clears the ground for sovereignty but does not occupy it. He is the instrument of restoration, not its beneficiary.

From an archetypal standpoint, this makes Hamlet a sacrificial hero rather than a dynastic one. He performs the purgative function of the royal hero by exposing corruption, destroying the usurper, and restoring the conditions of order; but he does so at the cost of his own claim. Authority survives him, but it does not belong to him.

Raglan's model thus clarifies something essential: Hamlet is not a failed hero because he does not become king; he is a transformed hero because kingship itself has become ethically compromised. To ascend the throne by repeating the logic of blood-vengeance would be to reinscribe the very disorder the play seeks to purge. Hamlet's refusal to seize power is therefore not weakness but discrimination. He recognizes that legitimacy cannot be restored merely by changing occupants; the mode of succession itself has become suspect.

This is why Fortinbras, not Hamlet, inherits Denmark. Fortinbras represents continuity of the old heroic order—decisive action, martial honor, dynastic claim—precisely because Hamlet has exhausted its moral credibility. The hero who undergoes the deepest ordeal cannot rule; the ruler who rules cannot undergo the ordeal. Shakespeare splits the archetype in two.

In Raglan's terms, then, *Hamlet* dramatizes a moment when the traditional fusion of heroism and kingship breaks apart. The hero still dies a meaningful death, but that death no longer consecrates his reign. It consecrates something more abstract and more modern: the principle that authority must be morally earned, not merely inherited or seized. Hamlet fulfills the function of the heroic archetype while surrendering its ancient prize.

This prepares the way for the next major reframing of Hamlet's heroism: not as failed succession, but as inverted quest—a journey that achieves its goal precisely by refusing its expected end. That inversion finds its most powerful articulation in Joseph Campbell's account of the monomyth.

The inverted quest (Campbell and the Monomyth)

Where Raglan reveals the political rupture in Hamlet's heroism, Joseph Campbell illuminates its narrative inversion. In *The Hero with a Thousand Faces,* Campbell argues that heroic stories across cultures share a common underlying structure, the monomyth or "hero's journey." The hero receives a call, resists it, encounters trials and helpers, descends into danger or death, gains a boon, and returns transformed to renew the community. Campbell's scheme is not a formula but a grammar, flexible enough to accommodate tragedy as well as romance.

Hamlet conforms to this grammar with uncanny precision, yet almost every stage is inflected with doubt, delay, or reversal. The result is not the absence of a heroic journey, but its tragic mutation.

The Call to Adventure arrives unmistakably in the form of the Ghost. Hamlet is summoned to avenge his father's murder and cleanse the kingdom. But unlike the mythic call, which often arrives with divine clarity, this call is morally unstable. The Ghost may be truthful or diabolical. Hamlet's world no longer guarantees that supernatural authority is benevolent. This uncertainty produces a prolonged Refusal of the Call, expressed not as cowardice but as epistemic caution. Hamlet insists on testing the spirit, staging the play, and securing certainty before action. The heroic journey stalls because knowledge, not courage, has become the prerequisite of legitimacy.

The Road of Trials unfolds not through physical combat but through psychological and ethical ordeals: feigned madness, surveillance, betrayal, the killing of Polonius, exile, and the constant pressure to perform roles that feel false. Denmark itself becomes what Campbell calls the "belly of the whale"—a closed, claustrophobic space of corruption from which escape seems impossible. Hamlet does not leave the ordinary world for a magical one; the ordinary world has itself become monstrous.

The traditional moment of Abyss or Death-and-Rebirth occurs during Hamlet's sea voyage to England. Sent away to be executed, he escapes through accident and piracy, emerging altered. When he returns in Act V, his language changes. The frantic oscillations of earlier acts give way to calm, acceptance, and a newly articulated sense of providence: "There's a divinity that shapes our ends." This is Hamlet's Apotheosis, but it is inward rather than triumphant. Illumination does not bring control; it brings surrender.

The Ultimate Boon Hamlet acquires is not kingship, power, or survival, but truth. Claudius is unmasked. Corruption is exposed. The moral reality of the court is forced into visibility. This boon is real, but it is also lethal. To carry it back intact would require Hamlet to live; and the monomyth's final stage, the Return, is where Shakespeare's inversion becomes complete.

Hamlet does not return. He dies.

Yet the community is renewed. Fortinbras inherits a purified kingdom. Horatio survives to tell the truth. The boon is transmitted, but not by the hero's continued presence. The heroic return is delegated. In Campbell's terms, the journey is completed not externally but symbolically: the hero's transformation enables the world's renewal even as it costs him his life.

This is why Hamlet can be described as a tragedy of the monomyth. Every stage is present, but each is ethically weighted, psychologically burdened, and finally inverted. Where the classical hero returns to rule, Hamlet returns to die. Where the mythic hero reconciles opposites, Hamlet accepts irreconcilability. Where the boon normally empowers the hero, here it consumes him.

Campbell himself observed that tragedy represents a variant of the hero's journey in which illumination arrives too late for personal salvation. Hamlet exemplifies this with exceptional clarity. The play insists that in a morally compromised world, heroic success may no longer be survivable. The journey remains necessary, but the price of completing it has changed.

Hamlet thus becomes a hero not despite his failure to return, but because of it. The monomyth is not abolished; it is internalized and moralized. The heroic victory occurs in consciousness, not in coronation. And this prepares the way for the final archetypal transformation of the chapter: Hamlet as the modern hero of individuation, whose struggle is not to conquer the world, but to separate ethically from the authority that commands him.

Individuation and the modern hero (Jung and Neumann)

If Campbell shows how *Hamlet* inverts the outward structure of the heroic quest, Carl Jung and, more decisively, Erich Neumann explain why that inversion was historically inevitable. In Jungian terms, Hamlet is not a hero who fails to act; he is a hero whose task has shifted from conquest to individuation, the painful separation of the conscious self from archaic authority, inherited commands, and unconscious compulsion.

Jung's conception of the hero archetype locates heroism within the psyche. The hero confronts the Shadow (those repressed aspects of the self projected outward), encounters the Anima (the soul-image mediating relation to meaning), and ultimately seeks the Self, a principle of wholeness that transcends ego and instinct. Applied to *Hamlet,* this framework immediately clarifies the play's distinctive gravity. Hamlet's enemies are not merely external. Claudius is not only a political usurper but a shadow-figure, someone who has acted out the very desires Hamlet finds abhorrent in himself. The disgust Hamlet feels toward Claudius is inseparable from self-recognition. The enemy "out there" mirrors a possibility "in here."

This is why Hamlet's struggle is so corrosive. The heroic demand to kill the king and take his place cannot be executed without psychic disintegration. To act too quickly would be to collapse into the Shadow, to become Claudius in another key. Jungian readings therefore interpret Hamlet's hesitation not as neurosis but as a desperate attempt to preserve psychic integrity in a world that rewards moral splitting.

The Ghost intensifies this conflict. In Jungian terms, the Ghost is not simply the father's spirit but an archetypal Father-command erupting from the unconscious—sacred, absolute, and terrifying. It speaks in the language of honor and vengeance, demanding obedience without reflection.

Hamlet's refusal to submit immediately is therefore not weakness but resistance to regression. He will not allow the archaic father-image to absorb his ego. He must test it, question it, and subject it to ethical scrutiny.

This insight finds its most precise articulation in *The Origins and History of Consciousness*. Neumann describes the heroic myth as the drama of consciousness separating itself from primordial authority—first from the devouring Mother, later from the commanding Father. In archaic myth, the hero proves himself by slaying the dragon or overthrowing the tyrant father. But in modernity, Neumann argues, the danger is no longer insufficient separation; it is premature identification. The modern hero must avoid collapsing into inherited forms of power that no longer correspond to ethical reality.

Hamlet exemplifies this transitional moment. He stands between two father-figures: the Ghost, representing sacred but archaic authority, and Claudius, representing illegitimate but effective power. To obey one blindly is to become monstrous; to tolerate the other is to accept corruption. Hamlet's task, therefore, is not simply to choose between fathers but to differentiate from both. He must act without becoming either the avenging warrior of the past or the calculating ruler of the present.

Seen this way, Hamlet's delay becomes the central act of individuation. He suspends action until he can act as himself, not as an instrument of inherited violence. He seeks a mode of action that does not annihilate conscience. This is why his final acceptance—"There's a divinity that shapes our ends"—marks not resignation but maturation. He relinquishes the fantasy of total control without surrendering moral agency. The ego steps aside for something larger, but it does not dissolve into obedience.

Neumann emphasizes that individuation often culminates not in triumph but in sacrifice. The hero who advances consciousness may not survive the transition he makes possible. Hamlet fits this pattern exactly. He achieves ethical clarity only when survival no longer matters. His death is not the failure of individuation but its completion. He acts at the moment when action no longer serves personal ambition, only truth.

This is what makes Hamlet a modern heroic archetype. Traditional heroes act first and understand later. Hamlet understands first and acts only when understanding becomes unbearable. His heroism lies in refusing to allow the old archetype to execute itself blindly. In Jungian–Neumannian terms, he is the hero who does not merely slay the father, but transcends the necessity of slaying him as a means of self-definition.

With this, the heroic archetype reaches a critical transformation. The hero no longer proves himself by ruling, conquering, or founding a dynasty. He proves himself by refusing illegitimate power, even when it is his by right, and by absorbing into himself the psychic cost of that refusal. Hamlet becomes the hero of conscience, the figure in whom myth turns inward and becomes ethics.

Robert Devereux, 2nd Earl of Essex by Marcus Gheeraerts the Younger

12. The Essex rebellion

The overmighty subject

Among the figures who orbit the late Elizabethan court, none burns more brightly or more destructively than Robert Devereux, 2nd Earl of Essex. His rise and fall are often treated as a cautionary tale of political rashness or personal vanity. Yet such explanations, while not false, remain inadequate. Essex is not merely a failed courtier. He is an archetype: the overmighty subject, the favorite who mistakes proximity to power for entitlement to it.

Essex possessed all the qualities that courts, and myths, reward early and punish late. He was young, striking, impetuous, and theatrically courageous. His military exploits, real and exaggerated, fed a popular image of heroic virtue; his intimacy with the queen fostered a sense of exceptionality that soon curdled into grievance. Elizabeth had elevated him, indulged him, and forgiven him repeatedly. But in doing so she had also placed him in the most dangerous of all positions: close enough to the throne to imagine himself indispensable, yet barred by law, age, and temperament from ever possessing it.

The Elizabethan polity had long experience with such figures. The overmighty subject is a recurring anxiety in English political thought, from the Wars of the Roses onward: a nobleman whose charisma, military reputation, or royal favor enables him to rival the crown itself. What distinguishes Essex is not that he occupied this role, but that he embraced it theatrically, even mythically. He did not simply wish to influence policy; he wished to embody national purpose. He conceived of himself as England's sword, its conscience, even its savior—an identity that left little room for obedience, patience, or institutional constraint.

This self-conception explains the paradox of Essex's rebellion. His 1601 rising was neither well planned nor realistically conceived. It relied on spectacle rather than strategy, on presumed affection rather than secured loyalty. He expected the city of London to rise spontaneously, the queen to be cowed by public sympathy, and the machinery of government to yield before his personal narrative. These were not the calculations of a Machiavellian schemer. They were the expectations of a man who believed his story had already been ratified by destiny.

Here the political merges with the theological. Essex's error was not simply ambition, but impatience with mediation. He could not tolerate the slow rhythms of counsel, law, and succession. He would not wait upon time, nor accept that favor was conditional. In this sense, his revolt belongs to a far older pattern than Elizabethan factionalism. It belongs to the myth of the radiant subordinate who rebels not because he lacks honor, but because he believes his honor entitles him to rule.

The tragedy of Essex lies precisely here. He was not base. He was excessive. He possessed genuine virtues—courage, generosity, magnetism—but these virtues, untempered by obedience, became destructive. His fall was not the exposure of hypocrisy, but the collision of heroic self-image with constitutional reality. England could admire him; it could not be governed by him. To recognize this is not to absolve Essex of responsibility, but to understand the depth of the crisis he precipitated. For when such a figure moves from grievance to action, the state is confronted with a choice as old as political order itself: whether charisma may override law, whether personal brilliance may displace institutional continuity, whether the favored angel may ascend without permission.

The Luciferian pattern

The Essex rising cannot be fully understood within the vocabulary of faction, policy, or personality alone. Beneath its political surface lies a far older structure, one that early modern readers, steeped in scripture and allegory, would have recognized instinctively. It is the pattern of the radiant subordinate who rebels against order itself, not out of malice, but out of pride and impatience.

In Christian theology, the fall of Lucifer is not precipitated by corruption or ignorance. He falls because he refuses to remain a minister. His cry, *non serviam*, is not a declaration of wickedness, but of autonomy. He will not wait, will not submit, will not accept mediation. The sin is not ambition in the abstract, but ambition that rejects time, hierarchy, and obedience. This distinction matters. Lucifer does not contest the existence of God; he contests the structure of authority. He believes his own brilliance entitles him to ascend without permission. His revolt is therefore symbolic before it is violent. The cosmic order is threatened not by brute force, but by charismatic exception, by the claim that one luminous figure may stand outside the law that governs all others.

This pattern maps with unsettling precision onto the career of Robert Devereux, 2nd Earl of Essex. Essex never denied the queen's sovereignty; he denied its conditions. He did not argue that Elizabeth should not rule, but that she ruled badly without him. His grievance was not exclusion from power, but insufficient recognition of his own indispensability. He had come to see himself not as a servant of the crown, but as its animating spirit, England's will made flesh.

The language of Essex's circle repeatedly drifts into this register. He is spoken of as the nation's champion, the sword of Protestant Europe, the man who ought to govern because he alone embodies courage and resolve. Such rhetoric is not yet treason, but it is already theological error. It confuses gift with office, virtue with authority, brilliance with legitimacy.

The crucial feature of the Luciferian pattern is haste. Lucifer will not wait for the fullness of time. Essex will not wait for counsel, reconciliation, or lawful succession. Delay feels to him like injustice. Mediation feels like insult. The machinery of governance, slow, impersonal, procedural, appears as an obstacle to destiny rather than its necessary form. This is why Essex's rebellion takes on a strangely theatrical quality. It is less a coup than a revelation meant to force recognition: a sudden appearance in the streets, an appeal to popular affection, an expectation that truth will assert itself spontaneously once unveiled. Like the fallen angel, Essex seems to believe that once he declares himself, the world must reorder itself accordingly.

In this light, the failure of the rising is inevitable. The Luciferian rebel always overestimates sympathy and underestimates structure. He believes that love can substitute for law, that admiration can replace obedience. When the city does not rise, when the institutions do not bend, the myth collapses—and collapse must follow swiftly, lest the contagion spread.

What is often forgotten is that early modern political thought regarded such moments with profound seriousness. A rebellion led by a charismatic favorite is not merely a challenge to a ruler; it is a challenge to order as such. It asserts that exception may rule, that brilliance may eclipse legitimacy, that the angel may become king by acclaim alone.

It is precisely at this juncture, when charisma threatens to supplant law, when personal narrative threatens institutional continuity, that the figure of the lawgiver becomes unavoidable. The state must answer not the man, but the principle he embodies. And it must do so even when the cost is moral anguish, public misunderstanding, and permanent reputational harm. That is the moment at which Francis Bacon is drawn inexorably into the drama.

Bacon's impossible position

Bacon's relationship with Essex had been real and complex. Essex had assisted him at moments when Bacon's career faltered; Bacon, in turn, had offered counsel that was often unwelcome precisely because it urged restraint, reconciliation, and patience. What divided them was not affection, but temperament. Essex believed himself born to command; Bacon believed power must be mediated by institution, time, and law.

By 1601, the machinery of the state was already in motion. Essex's actions were public, undeniable, and existentially threatening. Once armed force had been raised against the queen's authority, the matter passed irrevocably from the realm of personal relationship into that of constitutional survival. The prosecution of treason was not discretionary. It was the state asserting its right to continue existing.

Bacon did not design this confrontation or decide its outcome. He was summoned because he possessed something Essex conspicuously lacked: a disciplined understanding of law as an impersonal order, not a tool of faction or revenge. To refuse participation would not have preserved moral purity; it would have constituted an abdication of office. It would have signaled that private obligation could override public duty: precisely the principle Essex had already violated.

Here the mythic structure tightens. In the fallen angel narrative, the most painful role is not that of the rebel, but that of the loyal minister who must oppose him. The angel who remains obedient is condemned to appear cold, unfeeling, even cruel, because obedience lacks spectacle. Bacon's position is tragic in exactly this sense. He is required to speak the language of law against a man whose appeal is emotional, charismatic, and theatrical. The asymmetry guarantees reputational damage.

What is crucial, and rarely acknowledged, is that Bacon did not prosecute Essex as a personal enemy. He prosecuted an idea: that exceptional individuals may place themselves above order. His argument was not vindictive but structural. It rested on the premise that mercy without submission dissolves authority, and that favoritism, once armed, becomes indistinguishable from tyranny. This is why Bacon's conduct so unsettled later generations. He refuses the modern consolation of moral drama. He does not posture as tragic hero or conflicted friend. He speaks instead as the voice of continuity, of law as something older, colder, and more enduring than affection. In doing so, he sacrifices something no statute can restore: public sympathy.

The result was a wound that never healed. Bacon emerged intact in office, but damaged in narrative. He would carry the mark of this episode for centuries, while Essex, defeated, condemned, and executed, would accrue the melancholy glamour reserved for fallen favorites and romantic rebels. The paradox is stark: the man who preserved order was condemned as heartless; the man who endangered it was mourned as noble. Such reversals are not accidents of history. They are the predictable outcome of a culture that confuses charisma with virtue and rebellion with courage.

What disappears in this retelling is the original context of terror. Elizabethan England had vivid memories of civil war, dynastic bloodshed, and religious violence. The prospect of an armed favorite marching on London was not romantic; it was catastrophic. To prioritize institutional survival over personal loyalty in such circumstances was not cynicism, but prudence of the highest order. Yet prudence rarely survives translation across centuries. Each age projects its own anxieties backward, and the nineteenth century, confident in its constitutional stability, could afford to sentimentalize rebellion. It could admire Essex without fear of consequence, and condemn Bacon without risk. The moral economy had shifted.

The result is one of history's most durable inversions: the preservation of order becomes suspect, while its disruption becomes ennobled. Bacon's silence, discipline, and procedural rigor read as moral absence; Essex's impatience and defiance read as integrity. The lawgiver loses; the fallen angel gains. This inversion does not merely distort Bacon's character. It obscures a fundamental truth about political life: that civilizations are sustained not by the brilliance of exceptional individuals, but by the willingness of capable men to restrain themselves. When that restraint is mistaken for moral failure, history teaches precisely the wrong lesson. It is against this long shadow of misunderstanding that Bacon's reputation must be recovered, not by pleading his innocence, but by restoring the moral framework within which his actions were intelligible, and indeed necessary.

Every heroic pattern exacts a price. In the case of Francis Bacon, that price was neither execution nor exile, but something more insidious and enduring: the loss of moral sympathy in the imagination of posterity. The Essex affair marks the moment when Bacon assumes, fully and irrevocably, the role of the lawgiver, and accepts the wound that accompanies it. In myth and scripture alike, the lawgiver is rarely beloved. Moses does not enter the Promised Land. Solomon's wisdom curdles into suspicion. Michael, who casts down the rebel angel, inherits no songs of admiration. Order is preserved, but affection is forfeited. This is not an accident of narrative; it is a structural feature of moral life. Those who restrain power must absorb the resentment of those who prefer its spectacle. For Francis Bacon, the wound was both personal and symbolic. He survived the Essex crisis intact in office, but altered in perception. The charge that he had chosen advancement over loyalty would follow him long after the specific circumstances had faded from view. It became a lens through which all his later actions were interpreted, a ready-made explanation for every compromise, every fall, every human failing.

13. Bacon and Machiavelli

> I will never set politics against ethics; especially for that ethics are but a handmaid to divinity and religion.
>
> —Francis Bacon, *Considerations Touching a War with Spain*

Modern scholarship habitually pairs Francis Bacon with Niccolò Machiavelli as joint progenitors of modern political and intellectual realism, and this pairing is not without textual warrant. Bacon does cite Machiavelli on numerous occasions, often in a conciliatory or diplomatically neutral register, and he clearly regarded Machiavelli as a writer who could not be ignored by any serious observer of power. In this respect, the association is historically understandable. Machiavelli's decisive break with external authority, divine, natural, or customary, has come to function as a symbolic origin point for modern instrumental reason, and Bacon is frequently drawn into that lineage as a sympathetic heir. Yet this habitual alignment risks obscuring important distinctions. A closer examination of Bacon's writings reveals a persistent unease with precisely those Machiavellian commitments that have most shaped his reputation: the reversibility of virtue, the tactical manipulation of appearance, and the substitution of adaptability for moral continuity.

What complicates the picture further is that many of Bacon's engagements with Machiavelli turn out to be allusive references to a book known as *Anti-Machiavel*. Published anonymously in French at Geneva, *Anti-Machiavel* went through twenty-four printings in five languages and was quite influential in its time. It is attributed to Innocent Gentillet, a French Huguenot lawyer who fled to Geneva after the St. Bartholomew massacres in 1572, however, there is reason to doubt this attribution, and a bibliographer questioned it as early as 1584: "For my part, I believe that all these Gentillets are masks, and that the author of *Anti-Machiavel* is not known."[16]

I believe internal evidence suggests that Bacon himself wrote *Anti-Machiavel* while he was at Cambridge; this may sound absurd, but in over fifty places it parallels his later writings. I will refer to them as different authors, but my argument is that this is actually the early work of

---

[16] *Les bibliothèques françoises de La Croix du Maine et de Du Verdier*, volume I, p. 220. Paris: Saillant & Nyon, 1772.

Shakespeare. T. S. Eliot may have hinted about this, in the introduction he wrote for *The Wheel of Fire* (1949): "What at first appears to be their 'philosophy of life' sometimes turns out to be only a felicitous but shameless lifting of a passage from almost any author ... [Shakespeare] has his Montaigne, his Seneca, and his Machiavelli, or his Anti-Machiavelli like the others."

Beginning with the 1577 Latin translation, *Anti-Machiavel* bears a dedication to Francis Hastings and Edward Bacon, Francis Bacon's half-brother. The Bacon family's connections to Protestant Geneva went back to Lady Anne Bacon's father, Sir Anthony Cooke, who lived on the continent as a Protestant exile during the reign of Mary I; he corresponded with Calvin and met Theodore Beza, Calvin's successor in Geneva, who approved *Anti-Machiavel* for publication. The dedication exhorts Edward to

> Imitate the wisdom, sanctimony, and integrity of your Father, the right Honorable Lord Nicholas Bacon, Keeper of the broad Seal of England, a man right renowned; that you may lively express the image of your Father's virtues in the excellent towardness which you naturally have from your most virtuous Father.[17]

TO THE MOST FAMOVS YONG GENTLEMEN, AS WELL FOR RELIGION, MODESTIE, AND OTHER VERTVES, AS AL-ſo for kinred, *Francis Haſtings*, and *Edward Bacon*, moſt heartie ſalutations.

---

[17] The anonymous author of the *Arte of English Poesie* (1589), which bears a dedication to Francis Bacon's uncle Lord Burghley, was also intimately familiar with Sir Nicholas: "I have come to the Lord Keeper Sir Nicholas Bacon, and found him sitting in his gallery alone with the works of Quintilian before him; indeed, he was a most eloquent man, and of rare learning and wisdom, as ever I knew England to breed, and one that joyed as much in learned men and men of good wits."

The dedication opens with a story from Plutarch about the Greek statesman Solon talking with Thespis, the poet and actor for whom thespians are named, often called the inventor of tragedy:

> After Solon had seen Thespis' first edition and action of a tragedy, and meeting with him before the play, he asked if he was not ashamed to publish such feigned fables under so noble, yet a counterfeit personage. Thespis answered that it was no disgrace upon a stage, merrily and in sport, to say and do anything. Then Solon, striking hard upon the earth with his staff, replied thus: "Yea but shortly, we that now like and embrace this play, shall find it practiced in our contracts and common affairs." This man of deep understanding saw that public discipline and reformation of manners, attempted once in sport and jest, would soon quail; and corruption, at the beginning passing in play, would fall and end in earnest.

Bacon shared this concern with the pedagogy of the stage, as expressed in *The Advancement of Learning*:

> Dramatic poesy, which has the theatre for its world, would be of excellent use if well directed. For the stage is capable of no small influence both of discipline and of corruption. Now of corruptions in this kind we have enough; but the discipline has in our times been plainly neglected. And though in modern states play-acting is esteemed but as a toy, except when it is too satirical and biting, yet among the ancients it was used as a means of educating men's minds to virtue.

After the dedication, *Anti-Machiavel* opens with a discussion of inductive vs deductive reasoning:

> Aristotle and other philosophers teach us, and experience confirms, that there are two ways to come unto the knowledge of things. The one, when from the causes and maxims, men come to knowledge of the effects and consequences. The other, when contrary, by the effects and consequences we come to know the causes and maxims... The first of these ways is proper and peculiar unto the mathematicians, who teach the truth of their theorems and problems by their demonstrations drawn from maxims, which are common sentences allowed of themselves for true by the common sense and judgment of all men. The second way belongs to other sciences, as to natural philosophy, moral philosophy, physic, law, policy, and other sciences, whereof the knowledge proceeds more commonly by a resolute order of effects to their causes, and from particulars to general maxims.

This is very close to what is found decades later in the *Novum Organum*:

> There are and can be only two ways of searching into and discovering truth. The one flies from the senses and particulars to the most general axioms, and from these principles, the truth of which it takes for settled and immovable, proceeds to judgment and middle axioms . . . The other derives axioms from the senses and particulars, rising by a gradual and unbroken ascent, so that it arrives at the most general axioms last of all; which is the true but unattempted way.

Again and again, Bacon's reflections on constancy, reputation, suspicion, tyranny, and the moral dangers of simulation align more closely with Gentillet's anti-Machiavellian framework than with Machiavelli's own prescriptions. His insistence that virtue is constitutive rather than cosmetic, that power reveals rather than creates character, and that the erosion of moral continuity produces political instability, are all articulated most fully and explicitly within the anti-Machiavellian tradition. What *The Prince* authorizes is not merely dissimulation *in extremis,* but the institutionalization of dissimulation as a standing posture. Virtue becomes provisional, identity becomes reversible. The prince must appear "merciful, faithful, humane, religious, upright," yet remain inwardly prepared to become their opposites whenever necessity requires. This is not hypocrisy as weakness or lapse; it is hypocrisy as design. Virtue is stripped of intrinsic gravity and reduced to reputational utility.

It is here that Machiavellian realism crosses from prudence into something more radical: the liquidation of moral continuity as a governing constraint. Bacon's political and moral writings register a persistent unease with precisely this move. He allows for reserve, indirectness, prudential silence, and dissimulation at times; but he does not allow for the reversibility of moral identity. In Gentillet's *Anti-Machiavel,* Machiavelli is not rejected because he is cruel or cynical, but because he dissolves the conditions under which authority remains intelligible. Once virtue is reduced to appearance, the "credit of virtue" becomes actively dangerous: it enables exploitation without restraint. Constancy, by contrast, is treated not as one virtue among others, but as the condition of possibility for virtue as such. Without constancy, justice, mercy, and faith become interchangeable masks.

This is why both Bacon and Gentillet place such weight on constancy. Bacon writes that "constancy is the foundation on which virtues rest," a claim that closely echoes *Anti-Machiavel*: "constancy is a quality which ordinarily accompanies all other virtues; it is, as it were, of their substance and nature." This point is crucial. Constancy is not one virtue among others; it is the condition of possibility for virtue as such. Without it, mercy, justice, and faith are reduced to interchangeable masks—precisely the condition Machiavelli recommends. Power then becomes a theater of simulation rather than a moral office.

The same concern animates Bacon's reflections on office, suspicion, and reputation. In the essay "Of Great Place," he observes that "a place showeth the man, and it showeth some to the better, and some to the worse," a formulation that *Anti-Machiavel* renders proverbially: "honors change manners." Elevation does not create virtue or vice; it exposes what was already there. Rank dissolves disguise. The question is whether character can withstand visibility without corruption. Likewise, in the essay "Of Suspicion," Bacon notes that "men of base nature, once suspected, will never be true." Suspicion does not corrupt the virtuous; it crystallizes the corrupt. *Anti-Machiavel* makes the political analogue explicit: "the best fortress that is, is not to be thought evil by subjects; and if a prince is once thought so, there is no fortress that can save him." Reputation, once forfeited, becomes irreversible not because subjects are irrational, but because moral judgment precedes obedience. Where Machiavelli treats this as a problem of optics, Bacon and Gentillet treat it as a problem of being.

Shakespeare does not argue this thesis; he stages it. In *Measure for Measure*, a play obsessed with delegated, partial, and ideological authority, he writes "it is virtuous to be constant in any undertaking." Constancy here is what separates justice from zealotry, law from enforcement without restraint. In *The Two Gentlemen of Verona*, the line spoken by Proteus, "were man but constant, he were perfect" is devastatingly ironic. The character named for mutability confesses the very lack that ruins him. In Shakespeare's work, inconstancy is not cleverness or adaptability; it is the seed of betrayal, faction, and tyranny.

Anti-Machiavel gives this insight emblematic form. "As soon as the prince shall clothe himself with Proteus' garments," it warns, "and has no certitude in word or deed, men may well say that his malady is incurable, and that in all vices he has taken the nature of the chameleon." This passage does not merely describe Machiavellian behavior; it pathologizes it. Three things are happening at once. Proteus signifies radical mutability, the absence of stable form. The chameleon signifies adaptive appearance without inner substance. And the diagnosis of incurability marks this not as prudence under pressure, but as degeneration. Once identity itself becomes tactical, there is no longer a subject capable of rule—only a succession of masks. Shakespeare makes the same point with Gloucester (Richard III): when selfhood becomes simulation, tyranny is not an aberration but the logical outcome:

> I can add colours to the chameleon,
> Change shapes with Proteus for advantages,
> And set the murderous Machiavel to school.[18]

The pairing of Proteus and the chameleon was not idiosyncratic. It was a humanist commonplace, widely disseminated through Erasmus's *Adages*, where both figures function as warnings against inconstancy, unreliable counsel, and corrupted prudence. These were not obscure symbols but part of a shared moral lexicon that an educated reader would recognize immediately. This makes Francis Bacon's private indexing of the same imagery in his *Promus* collection especially revealing. Among its entries appears the compressed triad: "Chameleon, Proteus, Euripus." Euripus, the Aegean strait whose currents were thought to reverse direction predictably, completes the pattern. Taken together, the triad warns against policy without principle, knowledge without stability, and rule without orientation. It names not Machiavellian prudence, but the pathology Machiavellianism risks becoming.

Another parallel is found in Bacon's essay "Of Seditions and Troubles":

> Also, as Machiavel noteth well, when princes, that ought to be common parents, make themselves as a party, and lean to a side, it is as a boat that is overthrown by uneven weight on the one side... For when the authority of princes is made but an accessary to a cause, and that there be other bands that tie faster than the band of sovereignty, kings begin to be put almost out of possession.

Anti-Machiavel:

---

[18] *Henry VI Part 3*, act 3 scene 2

> For if he nourishes partialities among his subjects, he cannot possibly carry himself so equally towards both parties, but in them both will be jealousy and suspicion. Each party will esteem the other to be more favored, whereupon he will hate his prince, and by that means it may come to pass that the prince shall be hated by both parties; and so both the one and the other shall machinate his ruin, which he can hardly shun, having all their evil wills.

There are very clear Shakespearean dramatizations of exactly this insight. In fact, Shakespeare returns to it repeatedly, as if testing the same political theorem under different historical conditions. *King Lear* is the paradigmatic case, the purest dramatic embodiment of the principle that the ruler must be a common parent. Once he becomes partial, authority becomes "accessary to a cause"; factions form stronger bonds than sovereignty, the ruler is hated by all sides and displaced. Lear does not merely abdicate power; he redistributes it along factional lines, rewarding performative loyalty and punishing restraint. Once he aligns himself emotionally with Goneril and Regan, withdrawing from the role of impartial arbiter, he becomes a dependent partisan, a client rather than a sovereign, a king "almost out of possession." The result matches Anti-Machiavel line for line: jealousy between parties, mutual suspicion, hatred of the prince by both sides, and finally his political nullity. Lear's tragedy is not madness first—it is partiality first, madness second.

If *Lear* shows partiality through weakness, *Coriolanus* shows it through pride. Here the ruler (or ruling class) leans openly to one side: patricians vs. plebeians, honor vs. appetite, martial virtue vs. civic mediation. Rome ceases to function as a polity and becomes a battlefield of factions, each bound by loyalty stronger than the state itself. Coriolanus is destroyed not because he lacks virtue, but because he cannot occupy the role of common parent, he cannot speak to both sides, and thus becomes the symbol of factional domination. Rome, like Bacon's boat, capsizes from uneven weight.

One of the most revealing points of convergence between Bacon, *Anti-Machiavel*, and Shakespeare occurs around Machiavelli's appeal to the myth of Chiron. Machiavelli's claim that the education of princes requires learning how to act both as man and beast, lion and fox, became a kind of shorthand for the new political realism. Bacon addresses it directly, and with visible irritation, in *The Advancement of Learning*:

> In the fable that Achilles was brought up under Chiron the Centaur, who was part a man and part a beast: expounded ingeniously but corruptly by

> Machiavel, that it belongeth to the education and discipline of princes to know as well how to play the part of the lion in violence and the fox in guile, as of the man in virtue and justice.

The rebuke is carefully framed. Machiavelli's ingenuity is conceded, but it is labeled corrupt. What is at stake is not whether princes sometimes act forcefully or cunningly—Bacon is no naïf about power—but whether the myth authorizes a collapse of the human measure itself. That same objection is stated even more bluntly in *Anti-Machiavel*:

> But should we call this beastliness or malice, what Machiavelli says of Chiron? Or has he read that Chiron was both a man and a beast? Who has told him that he was delivered to Achilles to teach him that goodly knowledge to be both a man and a beast?

The force of the question lies in its refusal to accept Machiavelli's premise. The myth of Chiron does not teach oscillation between man and beast; it presupposes the distinction. To erase that distinction is not prudence but degradation. What Machiavelli redefines as wisdom, both Bacon and Gentillet identify as a loss of form.

This objection is not merely theoretical. It is dramatized with extraordinary clarity in *Timon of Athens,* in what is perhaps the most explicit repudiation of Machiavellian prudence in the Shakespearean canon:

> A beastly ambition, which the gods grant thee t'
> attain to! If thou wert the lion, the fox would
> beguile thee; if thou wert the lamb, the fox would
> eat thee; if thou wert the fox, the lion would
> suspect thee... What beast couldst thou be, that
> were not subject to a beast? And what a beast art
> thou already, that seest not thy loss in
> transformation!

The brilliance of this passage lies not merely in its imagery but in its logic. Shakespeare takes Machiavelli's lion–fox synthesis and runs it through a *reductio ad absurdum*. No beastly posture is stable. Each invites its own undoing. What Machiavelli presents as adaptive mastery is exposed as endless vulnerability. The problem is not choosing the wrong animal, the problem is becoming an animal at all. The closing line is decisive: "What a beast art thou already, that seest not thy loss in transformation!" This is almost a dramatic paraphrase of *Anti-Machiavel*'s central charge. Moral degradation is not a risk that accompanies Machiavellian prudence; it is its operating principle. Transformation itself is the loss. Timon's world, where loyalty is performative, friendship transactional, and power circulates through flattery and calculation, reveals the political consequence of this doctrine. Once politics is reduced to animal cunning and force, everyone becomes prey. That diagnosis aligns precisely with Bacon's fear that Machiavellian realism, once normalized, corrodes the conditions of civic trust rather than securing them.

Hamlet and *Anti-Machiavel*

Anti-Machiavel:

> When the emperor Claudius would espouse Agrippina, his brother's daughter, he made a law whereby he authorized the marriage of the uncle with the niece, which was published all over . . . indeed this marriage fell out not well for him; for Agrippina poisoned him to bring Nero to the empire, her son by another marriage; although Claudius had by his first wife Messalina a natural son called Brittanicus, whom Nero poisoned when he came to the empire. So that by the incestuous marriage wherewith Claudius had contaminated and poisoned his house, he and his natural son, who by reason should have been his successor, were killed with poison.

This passage is uncannily close to the moral architecture of *Hamlet*, and the proximity goes well beyond a shared classical anecdote. What Gentillet is doing here is supplying a ready-made moral template that Shakespeare appears to dramatize, rather than merely allude to.

In *Anti-Machiavel*, the story of the Roman emperor Claudius is not presented neutrally. It is a causal narrative: moral transgression precedes and generates political catastrophe. Law is bent to justify appetite; once that happens, the household becomes poisoned, literally and figuratively. This is precisely the logic Machiavelli suppresses and *Anti-Machiavel* insists upon. Shakespeare's Claudius reproduces this pattern with striking fidelity:

Anti-Machiavel	***Hamlet***
Incestuous marriage (uncle–niece)	Incestuous marriage (brother–sister-in-law)
Law altered to justify desire	Court rhetoric and "seeming virtue" justify the marriage

Anti-Machiavel	*Hamlet*
Poison as instrument of power	Poison as the means of regicide
Legitimate heir displaced (Britannicus)	Legitimate heir displaced (Hamlet)
Tyranny follows moral corruption	"Something is rotten in the state of Denmark"

Even the method, poison, carries over. In *Anti-Machiavel,* poison becomes the emblem of a house already morally infected. In *Hamlet,* the poison poured into the ear becomes the master symbol of corrupt persuasion, corrupted law, and corrupted lineage.

One of the most important elements in Gentillet's passage is this: *he made a law whereby he authorized the marriage...* This is not incidental. It is a direct indictment of Machiavellian statecraft, where law becomes an instrument of will rather than an expression of moral order. In *Hamlet,* Claudius does the same thing rhetorically rather than formally: he normalizes the marriage through court language, he reframes grief as political danger, and he substitutes policy for justice. Thus Claudius's Denmark, like Claudius's Rome, is a state where law no longer restrains desire but ratifies it. That is the precise nightmare *Anti-Machiavel* warns against.

Gentillet's conclusion is devastating: *by the incestuous marriage wherewith Claudius had contaminated and poisoned his house...* This is not metaphorical flourish, it is structural. The house (*domus*) is the seed of the state (*res publica*). Corrupt the one, and the other must follow. Shakespeare translates this directly into drama: the royal bed is corrupt, the court is corrupt, the state is corrupt, and nature itself is out of joint. Hamlet's famous intuition, that Denmark's sickness begins in the king's private crime, is exactly the lesson Gentillet extracts from Roman history.

Is this direct influence? What can be said confidently is that this episode from *Anti-Machiavel* provides a fully articulated anti-Machiavellian reading of Claudius; Shakespeare's *Hamlet* enacts the same moral logic, with the same crimes, symbols, and consequences; and the name Claudius itself functions as a signal, not a coincidence. Shakespeare does not merely borrow Roman history; he stages Gentillet's argument.

I should also briefly mention *The Great Assizes holden in Parnassus* (1645, attributed to George Wither), which features a court of poets and scholars, with Francis Bacon as Chancellor of Parnassus, before whom are arraigned authors charged with "strange abuses, committed against [Apollo] and the Nine Muses":

> He was accused, that he had used his skill,
> Parnassus with strange heresies to fill,
> And that he labour'd had for to bring in,
> Th' exploded doctrines of the Florentine,
> And taught that to dissemble and to lie,
> Were vital parts of humane policy.

"Th' exploded doctrines of the Florentine" can only refer to *Anti-Machiavel*. The court of Parnassus also includes William Shakespeare as "Writer of weekly accounts," Ben Jonson as "Keeper of the Trophonian Den," and the scholar Isaac Casaubon, a friend of Bacon's who was born in Geneva to Huguenot refugee parents. Casaubon is best known for proving that the *Corpus Hermeticum* dates from the Common Era; in a later chapter we will discuss Bacon's possible role in some of the Hermetic literature of his time. Bacon wrote in a letter to Casaubon: "To write at leisure that which is to be read at leisure matters little; but to bring about the better ordering of man's life and business, with all its troubles and difficulties, by the help of sound and true contemplations—this is the thing I am at."[19]

Machiavelli and *Leicester's Commonwealth*

Leicester's Commonwealth can be read as an English case study of the political pathology Gentillet anatomizes in the *Discours contre Machiavel*. Gentillet's polemic proceeds by exposing Machiavellian maxims in the abstract and then demonstrating their consequences for religion, justice, and sovereignty; the pamphlet reverses this procedure, beginning with concrete accusations against Robert Dudley and inviting the reader to infer the maxims that must govern a man who behaves in such a fashion. The author leaves little doubt as to the inference he expects, repeatedly identifying Dudley as one who acts by the counsel of "Seignior Machiavel my Lord's counsellor." The two works thus share not merely a moral outlook but a conceptual grammar.

---

[19] Spedding, James. *The Letters and the Life of Francis Bacon, Vol. IV*, p. 147. London: Longman, Green, Reader, and Dyer, 1868.

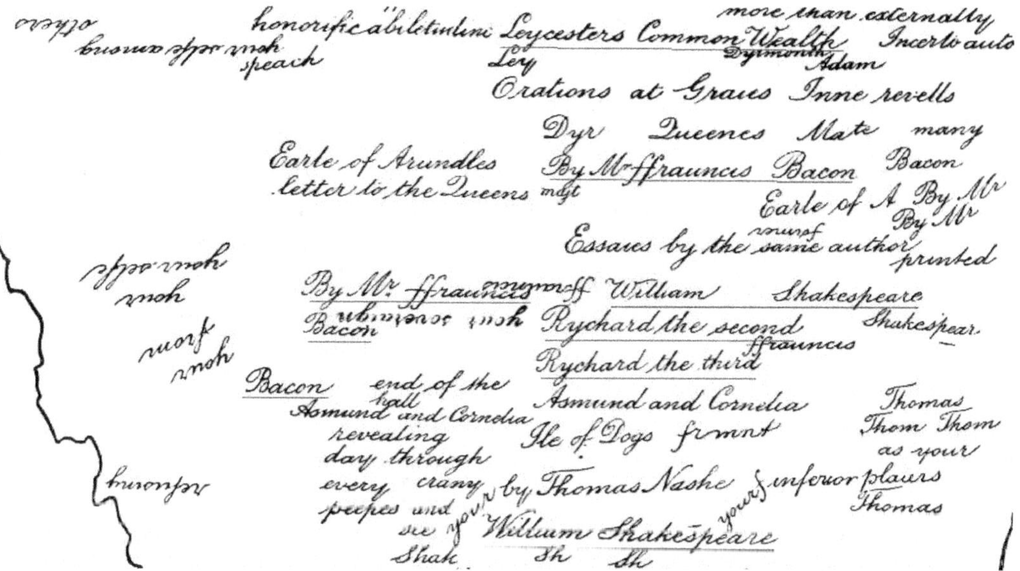
honorificabilitudini Leycesters Common Wealth more than externally Incerto auto
Ley Dymmoth Adam
Orations at Graies Inne revells
Dyr Queenes Mate many
Earle of Arundles letter to the Queens mate
By Mr ffrauncis Bacon Bacon
Earle of A By Mr
Essaies by the same author printed
By Mr ffrauncis William Shakespeare
Bacon Rychard the second Shakespear
Rychard the third ffrauncis
Bacon end of the hall Asmund and Cornelia Asmund and Cornelia
revealing day through every crany peepes and see your
Ile of Dogs frmnt
Thomas Thom Thom as your
by Thomas Nashe inferior plaiers Thomas
William Shakespeare
Shak Sh Sh

Northumberland Manuscript with "Francis Bacon," "William Shakespeare," "Leycester's Commonwealth"

Central to both texts is the insistence that Machiavellian power prefers secrecy to spectacle. Gentillet repeatedly condemns the use of hidden cruelty—poison, contrived accidents, silent removals—as the characteristic technique of the modern tyrant, precisely because it preserves outward reputation while securing inward fear. The pamphlet's obsessive return to allegations of poisoning and covert murder, beginning with the death of Amy Robsart, is therefore not incidental sensationalism but doctrinally precise. Leicester is depicted as one who obeys what the author calls "a settled rule of Machiavel which the Dudleys do observe, that where you have once done a great injury, there must you never forgive." Mercy would expose the crime; therefore the victim must be removed. What Gentillet theorizes as a structural danger to princely government, the pamphlet dramatizes as a pattern of lived practice, turning political life into a theater of suspicion.

Equally exact is the alignment on religion. Gentillet's most sustained objection to Machiavelli is not atheism in the crude sense, but the reduction of religion to an instrument of rule, a mask to be worn or discarded according to convenience. *Leicester's Commonwealth* repeatedly accuses Dudley of precisely this duplicity. His public Protestant zeal is portrayed not as conviction but as leverage: a means of intimidating rivals, dividing the realm, and consolidating personal authority. In this respect he exemplifies the Machiavellian "statist," and the pamphlet remarks with bitter irony that "when he plays the 'statist,' wringing very unluckily some of Machiavel's axioms to serve his purpose, then indeed he triumphs." Gentillet's warning that feigned piety corrodes religion rather than strengthening it is here translated into political narrative: instrumental godliness produces faction, hypocrisy, and ultimately impiety.

The pamphlet's handling of foreign policy likewise mirrors Gentillet's critique of Fortune. Gentillet insists that Machiavelli's exaltation of adaptability and audacity in place of providence encourages reckless opportunism, substituting moral restraint with tactical gambling. Leicester's intervention in the Low Countries is framed in exactly these terms: not as principled Protestant solidarity, but as ambition disguised as policy. Military incompetence and diplomatic confusion are attributed to a man who trusts stratagem over order, calculation over conscience. The pamphlet thus converts Gentillet's abstract claim, that politics severed from providence will collapse, into a concrete historical narrative of national risk.

Most important, however, is the shared deployment of the "evil counsellor" framework. Gentillet preserves the legitimacy of monarchy by locating corruption not in the prince but in those who govern the prince's will. Machiavellianism, in this account, does not seize the crown; it governs from behind the throne. *Leicester's Commonwealth* follows this logic with remarkable discipline. Elizabeth herself is repeatedly exonerated, even idealized, while Leicester is accused of monopolizing access, filtering information, and ruling without title. The implication is made explicit: "Why, then, the Earl of Leicester means and plots to become king himself," not by open usurpation, but by Machiavellian control of counsel. This is Gentillet's model almost verbatim: tyranny without a crown, power exercised invisibly through proximity rather than office.

The charges concerning justice and administration likewise track Gentillet's fear of legal formalism emptied of moral substance. Machiavelli, Gentillet argues, reduces law to an appearance maintained for convenience, subordinating justice to expediency and fear. Leicester is accused of manipulating courts, obstructing redress, and ensuring that law serves personal interest rather than public equity. The pamphlet identifies here a familiar Machiavellian sleight: "of driving men to attempt somewhat whereby they may incur danger or remain in perpetual suspicion or disgrace," so that enemies destroy themselves while the architect remains concealed. Law survives in form, but its spirit is hollowed out.

Economic exploitation plays a similar role. Gentillet criticizes Machiavelli for undervaluing the political danger of avarice, insisting that monopolies and oppression generate hatred that no amount of prudence can manage. Leicester's wealth is therefore presented not as evidence of success but as an index of systemic corruption. His enrichment widens the indictment from court politics to the commonwealth at large, supplying empirical confirmation of Gentillet's theoretical warning.

Underlying all these charges is a shared moral anthropology inherited from Erasmus: the Machiavellian as Proteus or chameleon, a figure without constancy who adapts endlessly because he possesses no stable moral core. Gentillet condemns such mutability as the negation of virtue; the pamphlet depicts Leicester as exactly this kind of man: one thing to one faction, another to the next, trusted by none because grounded in nothing. Mutability becomes not strength but moral vacancy, rendering governance impossible.

Read together, then, *Leicester's Commonwealth* emerges as the practical demonstration of Gentillet's anti-Machiavellian thesis. Every major accusation against Leicester corresponds to a Machiavellian maxim Gentillet condemns: secret cruelty, instrumental religion, reliance on Fortune, manipulation of law, enrichment at public expense, and the corrosion of counsel. The pamphlet does not merely attack a man; it stages an English experiment in Gentillet's political theology, warning that when cunning replaces virtue and prudence is severed from providence, corruption becomes systemic and ruin inevitable. In this sense, Leicester functions not simply as a historical villain, but as the English *Machiavel*—an embodied proof that Gentillet's fears were not speculative, but urgently real.

Vindiciae contra tyrannos

The same anti-Machiavellian logic governing constancy and virtue also underwrites the Protestant theory of resistance articulated in *Vindiciae contra tyrannos* (1579), translated into English as *A Defense of Liberty against Tyrants*. The work was published in Basel with a false imprint of Edinburgh, and issued pseudonymously as Stephanus Brutus Junius, a name deliberately invoking both Marcus Junius Brutus, assassin of Julius Caesar, and Lucius Junius Brutus, the legendary founder of the Roman Republic after the expulsion of the Tarquins—a genealogy of resistance that runs directly through Roman constitutional myth and into early modern political thought. Shakespeare portrayed the latter in *The Rape of Lucrece*—the beginning of the Roman Republic—and the former in *Julius Caesar*, the end of the Republic and beginning of the Roman Empire.

The choice of pseudonym is itself a rebuttal to Machiavelli, who notoriously warned that "Whoever takes up a tyranny and does not kill Brutus, and whoever makes a free state and does not kill the sons of Brutus, maintains himself for little time." Machiavelli treats Brutus as a structural threat to any regime, whether tyrannical or republican, a remainder that must be eliminated. The *Vindiciae*, by contrast, restores Brutus as a moral necessity. Resistance is not an accident of politics but its corrective principle when lawful authority dissolves into personal rule.

This contrast becomes explicit in the *Vindiciae*'s account of Tarquin the Proud, which defines tyranny not by cruelty alone but by the systematic abandonment of counsel, law, and covenant:

> Tarquinius Superbus was therefore esteemed a tyrant, because being chosen neither by the people nor the senate, he intruded himself into the kingdom only by force and usurpation . . . The true causes why Tarquinius was deposed, were because he altered the custom whereby the king was obliged to advise with the senate on all weighty affairs; that he made war and peace according to his own fancy; that he treated confederacies without demanding counsel and consent from the people or senate; that he violated the laws whereof he was made guardian; briefly that he made no reckoning to observe the contracts agreed between the former kings, and the nobility and people of Rome.

What is striking here is how closely tyranny is identified with inconstancy. Tarquin's crime is not merely usurpation at the outset, but the abandonment of stable forms: consultation, law, contract, and continuity. Kingship becomes will. Authority becomes improvisation. This is precisely the condition Bacon, Gentillet, and *Anti-Machiavel* identify as politically fatal. The parallel account in *Anti-Machiavel* reinforces the same diagnosis in more explicitly moral terms:

> Tarquin, who enterprised to slay his father-in-law king Servius Tullius to obtain the kingdom of Rome, showed well by that act and many others that he was a very tyrant. . . when he changed his just and royal domination into a tyrannical government, he became a contemner and despiser of all his subjects, both plebian and patrician. He brought a confusion and a corruption into justice; he took a greater number of servants into his guard than his predecessors had; he took away the authority from the Senate; moreover, he dispatched criminal and civil cases after his fancy, and not according to right; he cruelly punished those who complained of that

> change of estate as conspirators against him; he caused many great and notable persons to die secretly without any form of justice; he imposed tributes upon the people against the ancient form, to the impoverishment and oppression of some more than others; he had spies to discover what was said of him, and punished rigorously those who blamed either him or his government.

Here tyranny is rendered as a syndrome: concentration of force, liquidation of counsel, privatization of justice, surveillance, and punishment of dissent. What Machiavelli would redescribe as prudent security measures are reclassified as symptoms of moral degeneration. Crucially, this degeneration is not episodic but cumulative. Once Tarquin abandons lawful form, every subsequent act reinforces the disorder. Tyranny is thus not merely unjust, it is unstable.

This constellation of ideas passes almost verbatim into Shakespeare's poetic treatment of Roman history. The argument of *The Rape of Lucrece* opens with an account of Tarquin that closely echoes both *Anti-Machiavel* and the *Vindiciae,* and may exemplify what T. S. Eliot famously called Shakespeare's "shameless lifting" from Gentillet:

> Tarquinius, for his excessive pride surnamed Superbus, after he had caused his own father-in-law Servius Tullius to be cruelly murdered, and, contrary to the Roman laws and customs, not requiring or staying for the people's suffrages, had possessed himself of the kingdom . . . the people were so moved, that with one consent and a general acclamation the Tarquins were all exiled, and the state government changed from kings to consuls.

What matters is not the question of direct literary borrowing but the shared moral grammar. In all three accounts, *Vindiciae, Anti-Machiavel,* and Shakespeare, tyranny is defined by rupture: rupture with law, with counsel, with continuity, and with trust. The expulsion of the Tarquins is not framed as revolution but as restoration. Constancy, once broken at the top, can only be recovered by removal.

Seen in this light, Shakespeare's Roman poems do not merely recount classical history; they dramatize the same anti-Machiavellian principle we have traced in Bacon and Gentillet. When authority becomes Protean, answerable only to circumstance and advantage, it forfeits its claim to obedience. The Brutus Machiavelli fears is precisely the Brutus the anti-Machiavellian tradition requires: the figure who restores moral continuity by terminating simulation.

Anti-Machiavel in folio (1602)

The decision to print *Anti-Machiavel* in folio format in 1602 is a strong contextual marker of how the work was meant to function. Folio was not the default format for controversial or polemical writing; it was expensive, prestigious, and normally reserved for texts that aspired to authority: law books, classical histories, patristic works, Bibles, and major works of statecraft. To issue *Anti-Machiavel* in folio was to present it not as a pamphlet for debate, but as a reference work, something to be consulted, cited, and kept. That alone suggests confidence that the book could circulate openly among educated and official readers without fear of suppression.

The contrast with *The Prince* is striking. Although Machiavelli's text circulated widely on the Continent in Latin and Italian, it did not appear in English until 1640, nearly four decades later. This was not because it was unknown, but because it was regarded as dangerous, corrosive, and politically destabilizing. In Elizabethan and early Jacobean England, Machiavelli's ideas were already notorious—absorbed indirectly through rumor, drama, and polemic—but the text itself remained effectively *unlicensed*. That *Anti-Machiavel* could be printed openly, and in a format associated with institutional seriousness, while *The Prince* could not be printed in English at all, points to a clear asymmetry in what the English state was prepared to endorse.

Taken together, these facts suggest that *Anti-Machiavel* functioned as more than a private moral critique. It reads plausibly as a sanctioned counter-text, articulating an officially acceptable response to Machiavellian realism without naming it as policy. In a political culture deeply concerned with stability, succession, and the moral legitimacy of rule, Gentillet's work provided a way to confront Machiavelli's influence while reaffirming a Christian-humanist conception of sovereignty. The folio format reinforces this role: it presents the book as a serious intervention aligned with law, ethics, and governance, not as an underground provocation.

A
DISCOVRSE
VPON THE MEANES
OF VVEL GOVERNING AND

MAINTAINING IN GOOD
PEACE, A KINGDOME, OR
OTHER PRINCIPALITIE.

Divided into three parts, namely, The Counsell, the Religion, and the Policie, which a Prince ought to hold and follow.

Against *Nicholas Machiavell* the Florentine.

Translated into English by Simon Patericke.

LONDON,
Printed by Adam Islip.
1602.

The marginalization of Gentillet

Gentillet has, whether by design or by inertia, been effectively buried in modern scholarship. A revealing episode in this process is the 1974 Stewart and D'Andrea edition of *Anti-Machiavel,* whose preface is remarkable for its astonishingly dismissive tone. The editors remark that "we cannot expect today's scholars to concern themselves with such erudite puzzles," a statement that functions not as a gesture of humility but as an act of intellectual gatekeeping. It reads less as a neutral editorial judgment than as a warning: *do not look deeper*. It is historically unusual, indeed almost unprecedented, for the editors of a major scholarly reprint to discourage inquiry into the very complexities that justify republication. The effect is to close a line of questioning at precisely the point where it ought to begin.

Scholars have noticed this pattern, albeit quietly and without sustained follow-through. Harro Höpfl has observed that Gentillet is "persistently downplayed" despite his demonstrable historical impact. Sydney Anglo has gone further, lamenting that Gentillet's role in anti-Machiavellian discourse has been "systematically neglected." Paul A. Rahe has been more explicit still, calling Gentillet "the most underestimated political theorist of the sixteenth century." These assessments, emerging independently from different corners of early modern scholarship, suggest that the marginalization of Gentillet is not accidental. One need not posit deliberate suppression to recognize a recurring pattern of scholarly avoidance.

Why, then, would twentieth-century political theory find it convenient to sideline a major critic of Machiavelli? One reason lies in the dominant narrative constructed around Machiavelli himself. Modern political thought has largely rehabilitated Machiavelli as the founder of political realism, the harbinger of secular rationality, and even a proto-liberal or proto-republican thinker. This story requires that Machiavelli's opponents appear backward, dogmatic, moralistic, or naïve, figures who resisted modernity rather than helping to shape it. Gentillet is an awkward fit for this role. He is systematic rather than reactive, relatively secularized compared to earlier Huguenot writers, historically informed, and politically practical. His critique of tyranny overlaps in important ways with later constitutional and resistance theories. Rather than representing an obsolete moralism, Gentillet complicates the triumphalist narrative in which Machiavelli leads cleanly and unopposed into modernity. For that reason alone, he is easier to sideline than to integrate.

A second reason is that Gentillet opens the door to a set of textual and ideological puzzles that cross national, confessional, and even authorship boundaries—puzzles for which existing institutional frameworks offer no easy answers. Why does *Anti-Machiavel* anticipate themes, formulations, and anxieties that later appear with striking clarity in Shakespearean drama? Why are there strong echoes between Gentillet, the *Vindiciae contra tyrannos*, and early Stuart resistance theory? Why do English texts, including those associated with Bacon's circle, show signs of deep familiarity with Gentillet's arguments, despite modern British scholarship's reluctance to acknowledge such influence? And how did a French or Genevan polemic penetrate so thoroughly into Elizabethan intellectual life without leaving a neat documentary trail? Academic gatekeepers tend to be uncomfortable with questions that destabilize disciplinary boundaries or require speculative reconstruction across languages and genres. Faced with such puzzles, it is institutionally easier to say nothing at all. The Stewart and D'Andrea preface makes this avoidance almost explicit.

Nevertheless, several scholars have hinted that there is more to this story. Sydney Anglo has noted "striking and never-explained parallels" between Gentillet and English dramatic and political literature, suggesting the possibility of a lost or obscured channel of transmission. Höpfl has characterized the erasure of Gentillet as a matter of scholarly prejudice rather than historical fact. Rahe has described the paucity of Gentillet scholarship as "perplexing," given the work's documented circulation and influence. Jean-Louis Fournel and Jean-Claude Zancarini have argued that the traditional narrative surrounding Gentillet is incomplete and in need of revision. These are restrained, professional formulations, but taken together they amount to a shared intuition: something does not add up.

Is the modern burial of Gentillet therefore suspicious? Not in a conspiratorial sense, but in an ideological and institutional sense, very much so. Marginalizing Gentillet serves several functions at once. It protects the dominant modern reading of Machiavelli, avoids uncomfortable parallels with Shakespeare and English resistance theory, and keeps authorship debates confined within familiar and manageable boundaries. Shakespeare scholars, in particular, have little incentive to engage deeply with sixteenth-century Huguenot political philosophy, which threatens to complicate entrenched assumptions about intellectual lineage and national culture. Academia, like any institution, tends to bury what does not fit prevailing paradigms. Gentillet fits none of them, and so, for generations, he has been quietly set aside.

Authorship questions

In volume I of *Les bibliothèques françoises* by François Grudé de La Croix du Maine and Antoine du Verdier (1584), the entry on *Anti-Machiavel* includes the striking comment: « *Pour moy, je croy que tous ces Gentillets sont masques, & que l'autheur de l'Anti-Machiavel n'est point cogneu.* »—"For my part, I believe that all these Gentillets are masks, and that the author of the *Anti-Machiavel* is not known." This is not casual skepticism but a deliberate bibliographical judgment, offered by compilers whose very purpose was to identify, classify, and stabilize authorship.

What makes the statement especially important is its plural formulation "*tous ces Gentillets*." It implies that Innocent Gentillet was not merely suspected of being a pseudonym, but that the name itself may have functioned as a collective or reusable mask, a convenient authorial persona under which a politically dangerous work could circulate. In the context of late sixteenth-century Europe, where Machiavelli's name was radioactive, censorship severe, and the stakes of political theory extremely high, this is entirely plausible. The *Anti-Machiavel* presents itself as a French Protestant critique of Italian political immorality, yet its sophistication, breadth of sources, and dramaturgical awareness suggest a hand (or hands) operating at a higher strategic level than that of a single moralizing pamphleteer.

This testimony matters because it reopens the question of authorship at precisely the moment when the book is closest to state power. *Anti-Machiavel* appears in an expensive folio edition in 1602, decades before *The Prince* is printed in English, and long before Machiavelli is widely accessible to the English reading public. A contemporary French bibliographer's admission that the author is "not known" undermines later assumptions of transparent attribution and invites us to consider *Anti-Machiavel* as a programmatic intervention rather than a personal tract, an instrument rather than a confession.

Placed alongside Francis Bacon's lifelong engagement with Machiavelli, his insistence on masks and his own habit of indirect authorship, this remark from La Croix du Maine and Du Verdier becomes especially suggestive. At the very least, it licenses methodological caution: we are not obliged to accept "Gentillet" as a settled fact when early modern authorities themselves refused to do so. At most, it points toward a culture of strategic anonymity, in which political philosophy circulates under borrowed names, and where the true author remains concealed precisely because the work is meant to operate at the level of states, not reputations.

In short, this sentence from *Les bibliothèques françoises* is not marginalia. It is a contemporaneous acknowledgment that *Anti-Machiavel* already appeared, to informed readers, as a masked text, one whose intellectual provenance could not be straightforwardly named.

Bacon's position recovered

Seen in this light, Bacon's relation to Machiavelli is neither naïve nor deferential. His measured tone is method, not equivocation. He refuses to normalize hypocrisy as policy. Power, for Bacon, is a test of character, not a license to discard it. Constancy is not rigidity; it is orientation. Virtue is constitutive, not cosmetic.

What emerges across Gentillet, Bacon, and Shakespeare is not a doctrine but a shared moral grammar. Simulation proves unstable. Identity treated as tactic dissolves trust. Tyranny is not the exception; it is the endpoint. Modernity did not begin as a single, unified break with moral authority. It emerged as a struggle between competing visions of power—one instrumental, the other moral—whose consequences remain with us still.

14. Bacon and esotericism

> In truth, much of Bacon's life was passed in a visionary world, amidst things as strange as any that are described in the Arabian Tales, or in those romances on which the curate and barber of Don Quixote's village performed so cruel an *auto-da-fe*, amidst buildings more sumptuous than the palace of Aladdin, fountains more wonderful than the golden water of Parizade, conveyances more rapid than the hippogryph of Ruggiero, arms more formidable than the lance of Astolfo, remedies more efficacious than the balsam of Fierabras.
>
> —Thomas Babington Macaulay, "Lord Bacon"

Francis Bacon is often remembered as a champion of empirical science and a herald of modern rationalism, sometimes even as a secular prophet whose occasional references to God were little more than conventional gestures. That picture has proven increasingly difficult to sustain. Across his corpus, religious language is not peripheral or ornamental, but structural. Bacon consistently frames his project of the Great Instauration in theological terms, drawing on the biblical drama of Creation, Fall, and Restoration to articulate both the problem knowledge must address and the limits under which it must operate. The advancement of learning, for Bacon, was not an emancipation from religion but an act undertaken within it.

Bacon's understanding of the Fall is central. He believed that before humanity's lapse, Adam possessed both moral innocence and intellectual dominion over nature, gifts granted by God and lost through sin. Religion, in Bacon's account, could restore innocence; the arts and sciences could restore dominion. These were not competing paths but complementary ones. Knowledge pursued rightly was therefore a religious vocation, a form of charity toward mankind and reverence toward creation. At the same time, Bacon repeatedly warned that inquiry unguided by "sound religion" and moral restraint risked compounding the damage of the Fall rather than repairing it. Knowledge was powerful precisely because humanity was fallen.

This conviction underlies Bacon's adherence to the theology of the "two books": Scripture and Nature. Each reveals God, but each must be read according to its own rules. Bacon was explicit that confusing the domains of theology and natural philosophy deforms both. He opposed both superstition and atheism, arguing famously that shallow philosophy inclines the mind toward disbelief, while depth restores it to religion. His polemic was not against faith but against intellectual arrogance. True philosophy, he insisted, would ultimately confirm divine wisdom rather than undermine it.

These theological commitments complicate older interpretations that cast Bacon as a covert secularist using religious language instrumentally to advance a fundamentally materialist agenda. While such readings dominated much twentieth-century scholarship, they have been challenged by more recent work, particularly by Stephen A. McKnight and Steven Matthews. Their studies demonstrate that religious motifs are not rhetorical camouflage, but the animating core of Bacon's vision. McKnight shows that Bacon's idea of Instauration draws deeply on scriptural, apocalyptic, and soteriological themes, while Matthews emphasizes the breadth of Bacon's theological imagination, including his engagement with patristic and allegorical traditions alongside Protestant commitments. Together, they make it increasingly implausible to regard Bacon as an atheist or religious cynic.

Once Bacon's sincerity is granted, his attraction to esoteric modes of thought becomes more intelligible. Bacon was intrigued by the idea of a *prisca theologia,* an ancient wisdom transmitted through symbols, myths, and enigmas. In *The Wisdom of the Ancients,* he treats classical fables as repositories of veiled insight, deliberately encoded to exclude the unprepared. He explicitly endorses the use of esoteric writing to reserve deeper truths for "selected auditors." This is not occultism in the sensational sense, but a disciplined economy of disclosure. Bacon believed that truth could be dangerous if revealed without preparation, and that layered communication was sometimes ethically necessary.

This principle governs Bacon's own practice of writing. In "Of Simulation and Dissimulation" and *Valerius Terminus,* he distinguishes between what should be disclosed openly, what should be hinted at, and what should be withheld. Such distinctions align him with a long philosophical tradition of esoteric writing, not as secrecy for its own sake, but as restraint in a fallen world. Bacon was not concealing heretical doctrine; he was concealing power—specifically, the radical implications of a new science capable of transforming society before moral institutions were ready to govern it.

Seen whole, Bacon's thought occupies a liminal position between theology, esotericism, and emerging science. He was neither a secular technocrat nor an occult mystic, but a Christian reformer who understood that the recovery of knowledge after the Fall required law, restraint, and moral orientation. In this sense, Bacon's use of esoteric forms aligns with his broader role as a lawgiver: he sought to regulate not only institutions and methods, but the conditions under which truth itself could safely enter the world.

The image of Bacon that emerges is therefore neither reductive nor romantic. He was a thinker for whom knowledge was sacred, dangerous, and redemptive all at once. To sever the scientific from the theological in his work is not to modernize him, but to misunderstand the seriousness with which he confronted the birth of modern power. Bacon's enduring significance lies precisely in this tension: he believed humanity could regain something of Eden through knowledge, but only if knowledge itself were bound under law, charity, and reverence for the divine order it sought to comprehend.

The atheist caricature

> "May God, the Creator, Preserver, and Restorer of the universe, in accordance with his mercy and his loving-kindness towards men, protect and guide this work, both in its ascent to his glory and in its descent to the good of mankind, through his only Son, God with us."
>
> —Francis Bacon

An account of Francis Bacon's alleged atheism should probably begin with Thomas Hobbes, whose proximity to Bacon has long complicated the picture. Hobbes worked for Bacon in the 1620s, serving as a kind of intellectual assistant and amanuensis, and later emerged as one of the most formidable political philosophers of the seventeenth century. Hobbes's reputation as a materialist, and his reduction of politics and psychology to mechanistic principles, has occasionally been retrojected onto Bacon by association. Yet this move obscures more than it reveals. Hobbes's system represents not the fulfillment of Bacon's vision, but a radical narrowing of it: method without metaphysics, power without spiritual horizon. Where Bacon repeatedly insists on intellectual humility before both nature and God, Hobbes constructs a closed system in which fear, appetite, and motion suffice. The fact that Hobbes passed through Bacon's orbit tells us more about the fertility of Bacon's mind than about any shared metaphysical destination.

A similar truncation occurs in Bacon's reception by the French Enlightenment, where he was revered as a liberator of reason but stripped of his theological depth. Figures such as Voltaire hailed Bacon as the progenitor of experimental science and the enemy of scholastic obscurantism, a hero of intellectual emancipation whose authority could be marshaled against ecclesiastical power. In this context, Bacon became a symbolic weapon rather than a fully read author. The Enlightenment took what it required—empiricism, progress, utility—and quietly discarded what it found inconvenient: Bacon's insistence on the Fall, his Two Books doctrine, his repeated invocations of divine illumination, and his belief that knowledge divorced from moral and spiritual formation becomes dangerous rather than liberating. The result was a Bacon refashioned in the image of his admirers, a proto-positivist rather than a Christian reformer of knowledge.

This selective inheritance lies at the root of the modern atheist caricature. Bacon is often treated as if he initiated a purely secular project whose internal logic inevitably leads to disenchantment and technocracy. Yet such readings depend on ignoring large portions of his corpus, including his prayers, his meditations, and his insistence that true knowledge must proceed *in charity*. Bacon does not reject God; he rejects false confidence, whether theological or philosophical, that mistakes inherited opinion for truth. His war is not against mystery but against idolatry, including the idolatry of reason itself.

It is here that Bacon may best be understood, not as an atheist, nor as a heterodox subversive, but as what might be called a spiritual gnostic with a lowercase *g*. He clearly does not belong to the Gnostic traditions that posit a corrupt creation fashioned by a lesser deity; on the contrary, Bacon affirms the goodness of creation and the legitimacy of studying it. Yet he consistently treats knowledge as something that must be recovered, purified, and reintegrated after a primordial loss. His emphasis on illumination, ascent, and disciplined transformation aligns him more closely with traditions of spiritual knowledge than with modern secular rationalism.

The most direct contemporary testimony comes from William Rawley, who lived with Bacon during the last decade of his life and knew him not as a symbol but as a man. Rawley addressed the charge of atheism explicitly and without defensiveness: "This lord was religious; for though the world be apt to think great wits and politiques have a touch of the atheist, yet he was conversant with God." This is not the language of polite conformity or public piety, but of sustained inward practice. Rawley's remark suggests a Bacon whose intellectual daring did not estrange him from the divine, but rather compelled him toward it.

Seen in this light, the atheist caricature appears less as a discovery than as a symptom, a byproduct of modernity's discomfort with figures who refuse to separate knowledge, power, and spirituality into neatly policed domains. Bacon's tragedy may be that his most influential heirs adopted his tools while abandoning his discipline, and his method while forgetting his intention. To recover Bacon whole is not to deny the ambiguities of his legacy, but to recognize that the story of modern knowledge begins not with disbelief, but with a man who believed that truth, rightly pursued, was ultimately a form of devotion.

Esotericism, secrecy, and the governance of knowledge

The term esotericism carries a heavy burden of misunderstanding, and any serious discussion of its presence in early modern thought requires careful definition. In modern usage, the word often connotes occult belief, secret societies, or arcane doctrine accessible only to initiates. Here we use the term in a more precise and historically grounded sense: to describe modes of writing and transmission designed to regulate access to knowledge whose consequences were understood to be politically, morally, or socially dangerous.

In this respect, Francis Bacon's use of allegory, fable, aphorism, and selective disclosure belongs to a long and respectable philosophical tradition. Classical authors from Plato onward distinguished between teachings suitable for general audiences and those requiring preparation, discipline, or discretion. In the early modern period, this distinction was sharpened by new political pressures: religious persecution, censorship, factional conflict, and the emergence of knowledge capable of destabilizing established authorities. Writing could no longer be assumed to be a neutral act.

This problem has been most clearly articulated in modern scholarship by *Persecution and the Art of Writing*, in which Leo Strauss argued that many philosophers wrote in deliberately layered ways in response to environments hostile to free inquiry. Strauss's claim was not that such authors trafficked in secret doctrines for their own sake, but that they employed strategic indirection in order to protect both themselves and their readers. Philosophical writing under conditions of persecution required artifice, irony, and concealment—not to deceive, but to survive.

More recently, *Philosophy Between the Lines* by Arthur Melzer has expanded this insight, demonstrating that esoteric writing was not an occasional tactic but a widespread and systematic practice in pre-modern philosophy. Melzer identifies multiple reasons for esotericism beyond fear of persecution alone: the protection of fragile truths, the prevention of misuse by the unprepared, the preservation of social order, and the pedagogical need to provoke active rather than passive understanding. Crucially, Melzer emphasizes that esoteric writing often serves ethical restraint rather than elitist exclusion.

This framework provides a useful context for understanding Bacon's position. Bacon nowhere claims possession of hidden doctrines unavailable to others. On the contrary, he repeatedly insists that knowledge is to be advanced for the benefit of mankind. Yet he is equally insistent that knowledge must be governed—by method, by moral purpose, and by prudent disclosure. His distinction between "disclosed" and "enigmatical" writing in *The Advancement of Learning* is not a rhetorical flourish, but a response to the recognition that premature or indiscriminate dissemination can deform truth rather than propagate it.

Bacon's preference for fable, aphorism, and emblem in works such as *The Wisdom of the Ancients* reflects this concern. Myth functions there not as superstition, but as a medium capable of carrying insight without collapsing it into dogma or provoking immediate resistance. Allegory allows ideas to be contemplated obliquely, preserving both intellectual freedom and social stability. In this sense, Bacon's esotericism is inseparable from his role as lawgiver: concealment becomes a means of restraint, not of domination.

It is important to stress what I do not claim. Nothing in this study requires positing Bacon's membership in secret societies, adherence to occult doctrines, or participation in clandestine organizations. Nor does it depend on claims of concealed authorship. The concern throughout is structural and epistemic, not biographical. The question is not whether Bacon believed in esotericism as doctrine, but whether he employed esoteric forms in response to the dangers he foresaw in the expansion of human power.

Seen in this light, the parallels between Bacon and later figures associated with alchemical, Rosicrucian, or symbolic traditions need not be sensationalized. They reflect shared problems rather than shared institutions: how to reform knowledge without unleashing chaos; how to advance inquiry while preserving moral orientation; how to protect truth from both suppression and misuse. Bacon's solution was neither secrecy nor transparency in the modern sense, but regulated disclosure—a graduated revelation governed by prudence and ethical end.

This conception of esotericism also clarifies Bacon's repeated warnings against the confusion of distinct "lights." The danger he identifies is not knowledge itself, but knowledge misapplied—truth placed in the hands of appetite, ambition, or faction without adequate preparation. In this respect, Bacon anticipates later concerns about technological power outrunning ethical deliberation. His insistence on restraint, indirection, and law is not archaic, but prescient.

Finally, this clarification serves to distinguish Bacon's position from both naïve Enlightenment transparency and modern conspiracy-driven readings of esotericism. Bacon does not imagine a future in which all knowledge can be safely exposed without mediation, nor does he endorse hidden elites governing from behind the veil. He imagines instead a moral architecture for inquiry—one in which truth advances, but under law; power grows, but under restraint; and knowledge remains a trust rather than a possession.

Understood in this way, Bacon's use of esoteric forms ceases to be an embarrassment or anomaly. It becomes an integral part of his reforming project, consistent with his jurisprudence, his epistemology, and his moral vision. To ignore this dimension is not to render Bacon more rational or modern, but to misunderstand the depth of his responsibility at the moment when modern power was being born.

Bacon and esotericism

In "Actaeon and Pentheus, or a Curious Man" (*Wisdom of the Ancients*), Bacon frames fable-pair as a rebuke to the "curiosity of men in prying into secrets" with an "undiscreet desire" for "things forbidden." Pentheus, for Bacon, exemplifies the minds that "by the height of knowledge and nature in philosophy… with rash attempts, unmindful of their frailty, pry into the secrets of divine mysteries," and are punished with "perpetual inconstancy" and "wavering and perplexed conceits." The madness-image is not decorative: Pentheus "sees two suns," because he confuses (or cannot reconcile) "the light of nature" and "the light of grace." Bacon is not condemning knowledge as such; he's condemning knowledge that refuses its own limits—the attempt to make natural reason do the work of grace, or to force divine things into the categories of philosophical scrutiny. The result is not enlightenment but a kind of doubling, a fractured perception that spills into will and action: unresolved intentions, sudden passions, and restless oscillation.

Twelve Keys of Basil Valentine (1599)

Bacon discusses esotericism in several places. In a general sense, the essay "Of Simulation and Dissimulation" stresses the importance of knowing "what things are to be laid open, and what to be secreted, and what to be showed at half lights, and to whom and when." In *New Atlantis*, members of Salomon's House take an oath of secrecy regarding inventions and discoveries unfit for public consumption. In *The Advancement of Learning* Bacon distinguishes between "disclosed" (exoteric) and "enigmatical" (esoteric) writing, the latter allowing the author "to remove the vulgar capacities from being admitted to the secrets of knowledges, and to reserve them to selected auditors, or wits of such sharpness as can pierce the veil." In the same book he talks about obscurity among poets "when the secrets and mysteries of religion, policy, or philosophy are involved in fables or parables," a theme he explored in depth in *Wisdom of the Ancients*. In *Valerius Terminus* he again extols the practice of esoteric writing "both for the avoiding of abuse in the excluded, and the strengthening of affection in the admitted." In *De Augmentis*, Bacon differentiates two methods of communicating knowledge, the magistral and the initiative: the magistral from the initiative method of transmitting knowledge. "The first transmits knowledge to the crowd of learners, the other to the sons, as it were, of science. The end of the first is the acquisition of existing knowledge, of the second the creation of new knowledge."

Bacon's capacity for nuanced writing is shown in his response to Machiavelli. Exoterically, when Bacon alludes to Machiavelli his tone is of deference or respect; but if we dig deeper, most of these allusions are actually pointing to a book known as *Anti-Machiavel*, a vitriolic rebuttal to the Florentine originally published in French in 1576. In over fifty places, Bacon's writings parallel *Anti-Machiavel* closely; collectively, these allusions suggest that Bacon intended them to be discovered at a later date, and that he did not want his true opinion of Machiavelli known at the time. Possibly this is because Machiavelli was held in high regard by politicians and courtiers; *Anti-Machiavel* complains "he is of no reputation in the court of France who has not Machiavelli's writings at the fingers' ends, both in the Italian and French tongues, and can apply his precepts to all purposes, as the oracles of Apollo." Bacon may have thought adopting a hostile tone towards the Florentine would jeopardize the reception of his reform of knowledge and philosophy; for he also wrote "I like better that entry of truth which comes peaceably, as with chalk to mark up those minds which are capable to lodge and harbor such a guest, than that which forces its way with pugnacity and contention."

Another example of Bacon's remarkable capacity for esoteric writing is found in the *Novum Organum*:

> We ought to make a collection or particular history of all monsters and prodigious births or productions; and, in a word, of everything new, rare, and extraordinary in nature. But this must be done with the most severe scrutiny, lest we depart from truth. Above all, every relation must be

> considered as suspicious which depends in any degree upon religion, as the prodigies of Livy: and no less so everything that is to be found in the writers on natural magic or alchemy, or such authors who seem all of them to have an unconquerable appetite for falsehood and fable.

This is actually a complicated allusion to Rabelais' *Gargantua and Pantagruel*:

> I find by the ancient Historiographers and Poets, that divers have been born in this world after very strange manners, which would be too long to repeat; read therefore the seventh chapter of Pliny, if you have so much leisure: yet have you never heard of any so wonderful as that of Pantagurel . . . I pass by here the relation of how at every one of his meals he supped the milk of four thousand and six hundred Cows.

The parallels are labored: compare "particular history" and "Historiographers"; "prodigious births" and "born in this world after very strange manners"; "prodigies of Livy" and "seventh chapter of Pliny"; "unconquerable appetite" and "supped the milk of four thousand six hundred Cows." Bacon is taking pains to alert us to Rabelais, I believe, because he was then (as now) the most explicit major author with regard to alchemy. *Gargantua and Pantagruel* abounds with alchemical language and Rabelais introduces himself as the "abstractor of the quintessence"; the fifth book has a passage strongly suggestive of Bacon's reform of philosophy:

> 'Tis the Novelty of the Experiment, which makes Impressions of their conceptive, cogitative Faculties . . . Be Spectators and Auditors of every particular Phaenomenon, and every individual Proposition, within the extent of my Mansion, satiate your selves with all that can fall here under the Consideration of your Visual and Ausculating Powers, and thus emancipate yourselves from the Servitude of Crassous Ignorance.

Taken as a whole, Bacon's engagement with esoteric forms is best understood not as flirtation with secrecy, but as an extension of his lawgiving vocation. Esotericism, for Bacon, is not a claim to privileged access, but a strategy of governance. Powerful knowledge must be mediated, staged, and limited, not because truth is fragile, but because human judgment is. Myth, allegory, aphorism, and selective disclosure function as epistemic safeguards.

This perspective dissolves many false alternatives that have distorted Bacon's reception. He is neither a magician in disguise nor a cold rationalist stripping the world of meaning. He is a legislator of knowledge, seeking to bind inquiry under law before its consequences outrun human wisdom. The esoteric dimension of his writing is inseparable from this aim. It is not the residue of superstition, but a technology of restraint.

In this sense, Bacon stands closer to the great lawgivers of myth and history than to later Enlightenment optimists. Like Moses, he does not enter the promised land himself. Like Solon, he binds future generations and then withdraws. Like Prometheus, he brings fire, but insists that it be governed. The secrecy he employs is not concealment of truth, but protection of the future.

The Chymical Wedding of Christian Rosenkreutz

There is an old debate within Hindu philosophy concerning the relative superiority of bhakti yoga, the yoga of devotion or the heart, and jnana yoga, the yoga of knowledge or wisdom (pronounced *gnāna*, with a hard *g*). Devotion has usually been accorded pride of place: love, surrender, and grace appear warmer, more humane, and more spiritually attractive than intellectual discrimination. Yet jnana yoga has at least this much to be said in its favor: even if one falls short of full union with God, one does not fail altogether. Knowledge refines the mind, disciplines perception, and leaves the practitioner wiser, more capable, and more useful to the world.

A remarkable passage, often attributed to Francis Bacon, seems to display an implicit understanding of precisely this division of spiritual paths:

> By us doth the Bridegroom offer thee a choice between four ways, all of which, if thou dost not sink down in the way, can bring thee to his royal court. The first is short but dangerous, and one which will lead thee into rocky places, through which it will be scarcely possible to pass. The second is longer, and takes thee circuitously; it is plain and easy, if by the help of the Magnet thou turnest neither to left nor right. The third is that truly royal way which, through various pleasures and pageants of our King, affords thee a joyful journey; but this so far has scarcely been allotted to one in a thousand. By the fourth shall no man reach the place, because it is a consuming way, practicable only for incorruptible bodies.

This passage, found in *The Chymical Wedding of Christian Rosenkreutz* (1616), is among the strongest and most intriguing parallels to Indian spiritual classifications that one can cite. Unlike many Baconian attributions, it does not depend on fragile verbal resemblance or clever wordplay. It rests instead on something far more difficult to counterfeit: structural knowledge. What is being displayed here is an understanding of how spiritual paths are distinguished, ranked, and psychologically evaluated, not merely named.

The Hindu framing itself is well established. Classical Indian traditions, especially as later systematized in Yoga and Vedānta, distinguish among physical austerity (*tapas*), devotion (*bhakti*), knowledge (*jnana*), and selfless action (*karma*). Asceticism promises speed but courts danger; devotion is safer and more accessible, though slower; knowledge is rare, demanding, and royal; and certain transformative paths involving the body itself are reserved for the few and are perilous in the extreme. Crucially, jnana yoga is often regarded as inferior in warmth but superior in resilience: when devotion falters, knowledge still yields insight and benefit.

What makes the passage from the *Chymical Wedding* extraordinary is not the mere enumeration of "four ways," but the manner in which each is evaluated. The first path, associated with physical austerities, is described as short but dangerous, leading through rocky places scarcely passable. This is a textbook assessment of asceticism in Indian literature: rapid results, severe risks, bodily damage, and the temptation to spiritual pride. Medieval Christian writing, by contrast, tends far more often to glorify austerity than to warn so explicitly against it.

The second path is longer and circuitous but plain and easy, provided one follows the guidance of the "Magnet." This language closely resembles the ethos of bhakti: attraction rather than force, love rather than strain, progress guided by a pull toward the divine. The image of the magnet is especially telling. In devotional traditions, God draws the soul; salvation is less a conquest than a response. That imagery sits more comfortably within Indian devotion than within the dominant categories of scholastic Christianity.

The third path, explicitly called "the truly royal way," is the way of knowledge. It is joyful, adorned with pleasures and pageants, but reserved for the few: "scarcely allotted to one in a thousand." This description aligns uncannily with classical jnana yoga, which is repeatedly characterized as kingly, rare, demanding of inner maturity, luminous rather than punitive, and capable of producing joy rather than dryness. Most strikingly, knowledge here is neither cold nor merely corrective; it is celebratory. Medieval European thought typically opposed intellect to love or subordinated it to doctrinal correctness. This passage instead harmonizes insight and delight.

The fourth way is a consuming path of fire, practicable only for incorruptible bodies. Here the language turns alchemical, but the parallel extends further. Traditions of bodily transformation, whether in alchemy, Tantra, or Vajrayāna Buddhism, promise liberation through transmutation of the embodied self. They are also notorious for their dangers: failure is not merely disappointing but catastrophic. Once again, the evaluative framing is precise rather than rhetorical.

Such knowledge could, in principle, have reached Europe by the late sixteenth and early seventeenth centuries. Jesuit reports from India were circulating; Persian and Arabic traditions had already absorbed and reframed Indian yogic ideas; Neoplatonic, Hermetic, and Sufi currents had translated Eastern distinctions into Western idiom; and alchemy functioned as a universal symbolic language of transformation. Transmission is therefore possible, but what is implied here is not casual borrowing. It is deep synthesis.

Even setting aside questions of authorship, the significance for Bacon is clear. The passage reflects exactly the sort of mind Bacon possessed: comparative rather than parochial, taxonomic rather than dogmatic, pragmatic rather than pious. It is concerned with what works, not merely with what flatters orthodoxy. It ranks spiritual paths without condemning them and allows for partial success rather than total failure. If one falls short, one is still improved and useful. That attitude, knowledge as fruit-bearing, failure as progress, advancement without perfection, is quintessentially Baconian. In Hindu terms, it is precisely the defense of jnana yoga.

While it is tempting to attribute the resemblance between the *Chymical Wedding*'s "four ways" and Eastern yogic paths purely to universal archetypes, there is good reason to consider historical transmission routes. The parallels are too structurally intricate to be dismissed as mere coincidence. In fact, there is a long and complex history of philosophical and mystical ideas traveling between East and West.

THE

ALCHEMIST.

A

COMEDY

Acted in the Year 1610.

By the KINGS MAJESTY'S Servants.

With the Allowance of the Master of REVELS.

The Author B. J.

——*petere inde coronam,*
Unde priùs nulli velarint tempora Musæ. Lucret.

LONDON:

Printed and Sold by *H. Hills*, in *Black-Fryars*, near the *Water-side.*

Ben Jonson's *The Alchemist*

As early as the Presocratics, thinkers were influenced by ideas that had made their way from India through the Persian Empire. During the era when both Anatolia and India were part of the Persian Empire, cross-cultural exchange was not just possible but inevitable. Later, the Hellenistic period and the spread of Neoplatonism also saw Eastern metaphysical concepts weaving into the Western philosophical fabric.

The Crusades, too, facilitated a renewed exchange of ideas. Islamic scholars had preserved and expanded upon ancient philosophical traditions, and their works re-entered Europe along with a broader influx of mystical and hermetic knowledge. Hermeticism itself is a blend of Greek, Egyptian, and possibly even Eastern influences, and it became a vessel for transmitting and reshaping these ideas in the Renaissance.

In the case of Gurdjieff, his "Fourth Way" teachings are a synthesis drawn from Central Asian and Middle Eastern sources—regions where Eastern and Western mystical traditions intersected for centuries, particularly along trade routes like the Silk Road. Thus, it's not a stretch to see the "four ways" in the *Chymical Wedding* as part of a broader continuum of knowledge that traveled through these cultural arteries. Rather than emerging in isolation, these parallels likely reflect a rich history of transmission and adaptation. They show how spiritual and philosophical ideas have been exchanged, preserved, and transformed across continents and centuries.

The writings of Philalethes

The writings attributed to Eugenius and Eirenaeus Philalethes ("lover of truth") present the most sustained and sophisticated continuation of Baconian themes in the mid-seventeenth century. Whether authored by Thomas Vaughan or another figure moving within the same intellectual constellation, these texts repeatedly articulate positions that align not merely in language but in epistemic structure with Bacon's project.

In *Anthroposophia Theomagica, Aula Lucis,* and *Euphrates,* knowledge is framed as a process of recovery rather than invention, illumination rather than accumulation. Again and again, the emphasis falls on discipline, humility, and moral fitness. Knowledge is dangerous not because it is false, but because it is powerful. It must therefore be approached through purification of the knower as much as through manipulation of nature. This is Bacon's position stated in a different register.

Philalethes' invocation of Heraclitus is particularly revealing. Heraclitus' doctrine of hidden harmony and perpetual flux functions here not as metaphysical speculation, but as epistemic warning. Nature does not yield herself to force or haste; she must be courted, questioned, and obeyed. Bacon's insistence that nature must be "put to the question" lawfully, not tortured into confession, finds a clear analogue here. Inquiry is a moral act.

The critique of Aristotle that runs through the Philalethan corpus also mirrors Bacon's with remarkable precision. Aristotle is faulted not primarily for error, but for sterility. His philosophy is said to terminate in words rather than works, in systems rather than transformations. This is exactly Bacon's charge. The convergence suggests a shared reformist impulse rather than casual borrowing.

Philalethes' emphasis on inward transformation does not contradict Bacon's outward orientation toward works and institutions; it completes it. Bacon understood that no reform of method could succeed without reform of character. The Philalethan insistence on purification, illumination, and restraint supplies what Bacon often gestures toward but leaves implicit. The two together form a coherent vision: knowledge as lawful power exercised by disciplined agents for the common good.

What is most striking is how consistently these texts resist sensationalism. There is no promise of omnipotence, no exaltation of the knower as sovereign. Instead, knowledge is presented as service, obligation, and burden. This tone is unmistakably Baconian. It is also deeply unmodern.

Anthroposophia Theomagica features a title page echoing that of Bacon's *Novum Organum*, which also quotes Daniel 12:4: "Many shall go to and fro, and knowledge shall be increased." It echoes Bacon's criticisms of Aristotle:

> Aristotle is a poet in text; his principles are but fancies, and they stand more on our concessions than his bottom. Hence it is that his followers, notwithstanding the assistance of so many ages, can fetch nothing out of him but notions . . . their compositions are a mere tympany of terms. It is better than a fight in *Quixote* to observe what duels and digladiations they have about him.

Anima Magica Abscondita, again railing against Aristotle:

> Away then with this Peripatetical Philosophy, this vain babbling, as St Paul justly styles it . . . the spirit of error—which is Aristotle's—produceth naught but a multiplicity of notions . . . His followers refine the old notions but not the old creatures. And verily the mystery of their profession consists only in their terms. If their speculations were exposed to the world in a plain dress, their sense is so empty and shallow there is not any would acknowledge them for philosophers. In some discourses, I confess, they have Nature before them, but they go not the right way to apprehend her. They are still in chase but never overtake their game; for who is he amongst them whose knowledge is so entire and regular that he can justify his positions by practice.

ANTHROPOSOPHIA
THEOMAGICA:
Or
A Diſcourſe of the Nature of
Man and his ſtate after death;
Grounded on his Creator's Proto-
Chimiſtry, and verifi'd by a practicall
Examination of Principles in
the Great World.

By *Eugenius Philalethes*.

Dan:
Many ſhall run to and fro, and knowledge ſhall be increaſed.

Zoroaſter in Oracul.
Audi Ignis Vocem.

LONDON,
Printed by *T.W* For *H. Blunden* at the
Caſtle in *Corn-hill*. 1650.

The Vaughan–More pamphlet battle of the early 1650s reveals how Bacon's challenge to Aristotelian authority remained a live and contested force well after his death. In *Man-Mouse Taken in a Trap*, Eugenius Philalethes invokes Bacon not as a distant precursor but as an active moral standard: the champion of experience against inherited authority.

> Had Bacon liv'd in this unknowing Age,
> And seen Experience laugh'd at on the Stage,
> What Tempests would have risen in his Blood
> To side an Art, which Nature hath made Good?
> …
> Tell me in earnest, dost thou think tis fit
> To believe all that Aristotle writ?
> Though he was blinded, yet experience can
> Sever the clouds, and make a clearer man.

The verses explicitly rehearse Bacon's core argument—that Aristotle's dominance rests on unexamined reverence rather than tested truth, and that experience, not metaphysical speculation, is the proper means of clearing intellectual "clouds." What is striking is that this appeal is directed not merely against scholasticism but against Cambridge Platonism itself, suggesting that by mid-century Platonism had assumed a conservative role, defending order and abstraction against experimental inquiry.

> The second project is to be more learned and knowing than Aristotle, that great Light (as thou doest blindly all him) of these European parts for these many hundred years together: and not only so, but to be so far above him that I may be his master, that I may lug him and lash him, as Harry Moore's breech should be lash'd. Pish! here is a project indeed, to do all this is nothing.

The satirical excess of Vaughan's rhetoric, imagining Aristotle dragged, mastered, and flogged, should not be read as mere bravado, but as a pointed exposure of the absurdity of treating ancient authority as sacrosanct. The controversy thus demonstrates that Bacon's revolt was not a short-lived methodological correction but a civilizational rupture whose aftershocks were felt precisely because institutional philosophy resisted it. In this sense, Vaughan stands as a Baconian inheritor, preserving experimental rigor where universities increasingly substituted metaphysical reassurance for inquiry.

Euphrates again sounds very much like an unrestrained Francis Bacon attacking Aristotle:

> I have often wondered that any sober spirits can think Aristotle's philosophy perfect when it consists in mere words without any further effects; for of a truth the falsity and insufficiency of a mere notional knowledge is so apparent that no wise man will assert it… did not Aristotle's science—if he had any—arise from particulars, or did it descend immediately from universals? . . . I have learned long ago, not from Aristotle but from Roger Bacon, that generals are of small value, nor fitting to be followed, save by reason of particulars. And this is evident in all practices and professions that conduce anything to the benefit of man.

In *Aula Lucis*, the speaker declares an explicit aversion to "barren" knowledge and frames true understanding in terms of fertility, generation, and performance:

> I could never affect anything that was barren, for sterility and love are inconsistent. Give me a knowledge that is fertile in performances, for theories without their effects are but nothings in the dress of things.

This is not decorative rhetoric. It encodes a precise epistemological criterion: knowledge proves itself only by what it produces. Abstraction without effect is not merely incomplete; it is ontologically thin—mere appearance. The language of love, generation, and fruitfulness places knowing within a quasi-biological, even sacramental economy, where truth must issue forth into works.

Bacon, *Of the Interpretation of Nature*:

> Knowledge that tendeth but to satisfaction is but as a courtesan, which is for pleasure and not for fruit or generation.

Bacon, *The Great Instauration*:

> That wisdom which we have derived principally from the Greeks is but like the boyhood of knowledge, and has the characteristic property of boys: it can talk, but it cannot generate; for it is fruitful of controversies but barren of works.

Bacon's formulations in *Of the Interpretation of Nature* and *The Great Instauration* articulate the same judgment with strikingly similar metaphors. Knowledge sought only for "satisfaction" is likened to a courtesan—pleasurable, alluring, but sterile—while Greek-derived wisdom is dismissed as the boyhood of knowledge, talkative yet impotent, "fruitful of controversies but barren of works." Across all three texts, sterility is not a neutral defect but a moral and philosophical failure. What matters is not discourse, system, or elegance, but generation: works, effects, transformations in nature and in human power over nature.

Taken together, these passages show that the author of *Aula Lucis* was not merely echoing Bacon stylistically but internalizing his central evaluative axis. The opposition between barren theory and fruitful knowledge had become a shared criterion by which intellectual legitimacy was judged, especially in alchemical and experimental circles resistant to university metaphysics. Bacon's revolution thus persists not only in method but in metaphor: knowledge must be generative, productive, and incarnate in effects, or it is no knowledge at all.

One passage in *Aula Lucis* has several allusions to Bacon: his title and rumored royal descent ("noble Verulam"); his heraldic motto, *mediocria firma* (the middle ground is firm), and his phrase for conveying secret knowledge in text, *traditio lampadis*:

> Had their doctrine been such as the universities profess now, their silence indeed had been a virtue; but their positions were not mere noise and notion. They were most deep experimental secrets, and those of infinite use and benefit. Such a tradition then as theirs may wear the style of the noble Verulam and is most justly called a Tradition of the Lamp . . . yet I cannot deny that some of them have rather buried the truth than dressed it. For my own part, I shall observe a middle way, neither too obscure nor too open, but such as may serve posterity and add some splendor to the science itself.

This passage is one of the clearest acknowledgments that Bacon was not merely an inspiration, he was a model of intellectual self-consciousness for later experimental and esoteric writers. The reference to "the noble Verulam" simultaneously invokes Francis Bacon by title and gestures toward the long-circulating rumor of his royal descent, subtly reinforcing Bacon's own cultivated image as both aristocratic lawgiver and philosophical founder. More than flattery, this establishes authority: the "Tradition of the Lamp" (*traditio lampadis*) is explicitly aligned with Bacon's conception of knowledge as something handed forward across generations, not hoarded, extinguished, or dissolved into mere speculation.

Equally revealing is the author's deliberate appeal to Bacon's heraldic and methodological ideal of the middle way—*mediocria firma*. The writer criticizes earlier practitioners for having "buried the truth rather than dressed it," a complaint Bacon himself frequently levels against both scholastics and obscurantists. The proposed alternative, "neither too obscure nor too open," is unmistakably Baconian: a controlled transparency, sufficient to serve posterity while guarding against vulgar misunderstanding or misuse. This is precisely the balance Bacon sought in his own symbolic writing, fables, aphorisms, and methodological fragments. The passage thus shows Bacon functioning not only as a philosophical authority but as a template for authorial posture: how to transmit dangerous or powerful knowledge responsibly, how to combine secrecy with public utility, and how to add "splendor to the science itself" without sacrificing fertility or truth. In short, *Aula Lucis* is not borrowing Baconian language accidentally—it is consciously situating itself within Bacon's unfinished project.

Bacon frequently quotes the aphorism of Heraclitus, "dry light is best soul";

Aula Lucis:

> Hence it is that I move in the sphere of generation and fall short of that test of Heraclitus: "Dry light is best soul."

Wisdom of the Ancients:

> It was excellently said by Heraclitus, "A dry light makes the best soul."

Novum Organum:

> The human understanding resembles not a dry light, but admits a tincture of the will and passions which generate their own system accordingly.

Bacon, "Of Friendship":

> It is in truth of operation upon a man's mind, of like virtue as the alchymists use to attribute to their stone for man's body; that it worketh all contrary effects, but still to the good and benefit of nature. But yet without praying in aid of alchmyists, there is manifest image of this in the ordinary course of nature . . . Heraclitus saith well in one of his enigmas, "Dry light is ever the best." And certain it is, that the light that a man receiveth by counsel from another, is drier and purer than that which cometh from his own understanding and judgment; which is ever infused and drenched in his affections and customs.

~

Aula Lucis:

> It is my design to make over my reputation to a better age, for in this I would not enjoy it, because I know not any from whom I would receive it.

Bacon, last will:

> For my name and memory, I leave it to men's charitable speeches, to foreign nations, and the next ages.

~

Aula Lucis:

> Future times, wearied with the vanities of the present, will perhaps seek after the truth and gladly entertain it. Thus you will see what readers I have predestined for myself.

Bacon, *Valerius Terminus*:

> Publishing in a manner whereby it shall not be to the capacity nor taste of all, but shall as it were single out and adopt his reader, is not to be laid aside, both for the avoiding of abuse in the excluded, and the strengthening of affection in the admitted.

~

Advancement of Learning:

> Let no man upon a weak conceit of sobriety or an ill-applied moderation think or maintain that a man can search too far, or be too well studied in the book of God's word, or in the book of God's works, divinity or philosophy; but rather let men endeavour an endless progress or proficience in both; only let men beware that they apply both to charity, and not to swelling; to use, and not to ostentation; and again, that they do not unwisely mingle or confound these learnings together.

Euphrates:

> Surely I am one that thinks very honourably of Nature, and if I avoid such disputes as these it is because I would not offend weak consciences. For there are a people who though they dare not think the majesty of God was diminished in that He made the world, yet they dare think the majesty of His Word is much vilified if it be applied to what He hath made—an opinion truly that caries in it a most dangerous blasphemy, namely, that God's Word and God's work should be such different things that the one must needs disgrace the other.

This pairing is exceptionally revealing, because it shows not merely verbal affinity but a shared theological settlement of the Baconian problem: how to pursue experimental knowledge without impiety, and how to defend the study of nature against religious suspicion.

In *The Advancement of Learning*, Francis Bacon articulates his doctrine of the two books: the book of God's Word and the book of God's Works. Bacon rejects the notion that inquiry can go "too far" in either domain, insisting instead on "endless progress or proficience" in both divinity and philosophy. The only cautions he allows are moral and practical, not epistemic: knowledge must be directed to charity rather than swelling, use rather than ostentation, and the two domains must not be confounded, even as both are legitimately pursued. Crucially, Bacon is not erecting a wall between theology and natural philosophy; he is regulating their *relation*, so that each may flourish without corruption.

The passage from Euphrates advances the same settlement, but in a more polemical register. The speaker defends the study of nature as an act of reverence, not presumption, and identifies the true danger not in experimental inquiry but in the belief that applying Scripture to nature somehow "vilifies" God's Word. That belief, he argues, implies an implicit blasphemy: that God's Word and God's Work are in tension, such that honoring one must diminish the other. This is precisely the anxiety Bacon sought to defuse decades earlier. Where Bacon speaks as a lawgiver of learning, Euphrates speaks as a defender against religious scruple; but both insist on the same principle, that God cannot be divided against Himself, and that a right reading of nature is not a rival to revelation but its companion.

Taken together, these passages show how thoroughly Bacon's theological framework had been absorbed by later experimental and esoteric writers. The concern is no longer whether natural philosophy is permissible, but how to conduct it without provoking weak consciences, how to preserve piety while extending knowledge. What Bacon formulates as a programmatic moderation—progress without pride, distinction without separation—reappears in *Euphrates* as a defense against a new kind of reactionary fear. The continuity suggests that Bacon's most durable legacy was not merely methodological, but theological: the conviction that truth in nature and truth in Scripture are mutually reinforcing, and that to oppose them is the real irreligion.

Eugenius Philalethes frequently quotes "the divine Virgil," "who was a great poet but a greater philosopher." Francis Bacon felt the same, citing "the best poet [known] to the memory of man" more than any other author—but usually in a scientific or philosophical context. Bacon wrote "certain critics are used to say hyperbolically, that if all sciences were lost, they might be found in Virgil"; *Anti-Machiavel* avers:

> If our youths gave themselves only to Virgil to learn all Latin poetry, it is enough; and that author alone, compared to whom all others are but small rivers, might teach them all the poetry that need be known... he who well understands Virgil has no need of others for the understanding of poetry. And in every science it seems to be the best, that men may well employ their time, which is dear and short, to read few books, to make good choice of them, and to understand them well.

This sounds like Bacon's famous aphorism: "Some books are to be tasted, others to be swallowed, and some few to be chewed and digested." This cluster of references is important because it shows Virgil functioning as a shared philosophical authority, not merely as a poetic ornament, across Baconian, anti-Machiavellian, and Philalethan circles. For Eugenius Philalethes, the repeated invocation of "the divine Virgil"—"a great poet but a greater philosopher"—signals an understanding of poetry as a vehicle of veiled natural and moral knowledge. Virgil is treated not as a stylist to be imitated, but as a thinker whose poetry encodes truths about nature, order, generation, and empire. This attitude aligns closely with Francis Bacon, who cites Virgil more frequently than any other author, and almost always in a scientific, moral, or philosophical register. Bacon's remark that "certain critics are used to say hyperbolically, that if all sciences were lost, they might be found in Virgil" is not casual praise: it reflects Bacon's conviction that ancient poets, Virgil above all, encoded observations about nature and human affairs in imaginative form, anticipating later systematic inquiry.

What sharpens the point is the convergence with *Anti-Machiavel*, which elevates Virgil to near exclusivity: if youth were to study only Virgil, *"it is* enough," for he alone contains all the poetry worth knowing. The language is telling—other poets are "small rivers," Virgil the source. This is not aesthetic maximalism but moral and civil pedagogy: Virgil teaches how power, virtue, fate, and labor interact, which is precisely what Machiavellian writers reduce to technique. Thus, in *Anti-Machiavel*, Virgil becomes the antidote to Machiavelli: a poet who reveals the tragic costs of empire, the limits of force, and the necessity of pietas.

Taken together, these strands suggest that Philalethes' reverence for Virgil is not incidental but inherited through a Baconian and anti-

Machiavellian valuation of poetic wisdom. Virgil stands as a figure who unites poetry, philosophy, and natural insight—someone whose work exemplifies Bacon's belief that myth and verse can preserve truths later unfolded by method, and Gentillet's insistence that moral formation precedes political cunning. In this light, Virgil is not merely "the best poet," but a guardian of integrated knowledge, bridging imagination, nature, and ethical order in a way that experimental philosophy sought not to abolish, but to complete.

EVPHRATES,

OR THE

WATERS

OF THE

EAST;

Being a ſhort Diſcourſe of that *Secret Fountain*, whoſe *Water* flows from *Fire*; and carries in it the *Beams* of the *Sun* and *Moon*.

By *Eugenius Philalethes.*

Sadith ex Lib. Sacro.

Et dixit Deus, cujus Nomen ſanctifice-tur: Fecimus ex Aqua omnem Rem.

London printed for *Humphrey Moſeley* at the Princes Arms in St. *Paul's* Church-yard 1655.

Again, *Euphrates* sounds very much like Bacon:

> Before his Fall man was a glorious creature, having received from God immortality and perfect knowledge; but in and after his Fall he exchanged immortality for death and knowledge for ignorance.

The theological framework articulated in *Euphrates* closely mirrors that of Francis Bacon, not merely in language but in underlying structure. Both texts begin from a shared postlapsarian anthropology: before the Fall, humanity possessed immortality and perfect knowledge; afterward, it inherited death and ignorance. In Bacon's thought, this loss is not only moral but epistemic, and the sciences are justified as a partial, disciplined attempt to repair what was lost in Adam. *Euphrates* expresses the same premise with striking directness, treating ignorance not as humility but as the consequence of sin. In both cases, knowledge is not opposed to piety; it is a legitimate, even necessary response to the Fall, provided it is pursued within proper moral and theological bounds.

After Thomas Vaughan purportedly died from alchemical experiments in 1666, another series of tracts were issued under the name Eirenaeus Philalethes. Professor Steven Mathews has argued recently that Irenaeus of Lyon was a significant influence on Bacon; the first of these alchemy texts went out under "Eirenaeus Philoponus Philalethes," which has additional significance for Bacon in the sense that John Philoponus was the most formidable anti-Aristotelian of late antiquity, remembered for demonstrating—centuries before Bacon—that Aristotle could be wrong about motion, matter, and time itself. The pseudonym quietly situates Philalethes within a submerged lineage of anti-Aristotelian natural philosophy that had long survived outside the schools, awaiting institutional vindication. Additionally, Irenaeus of Lyon is known for his writings against the Gnostics, including the followers of Basilides and Valentinus. These names were revived in Bacon's time as the alchemical writer "Basil Valentine"; possibly he was trying to revive a kind of debate similar to what happened in the early years of Christianity.

Benedictus Figulus, *Pandora magnalium naturalium aurea et benedicta de Paracelse* (Strasbourg, 1608) also attacks Aristotle, again sounding very much like a relaxed Bacon:

> When reviewing the whole course of my studies, from my youth up, I find—and have indeed hitherto found in my work, and clearly experienced more and more with the lapse of time, as daily experience shows is wont to happen to the true believer and right naturalist—that there are three kinds of Philosophy or Wisdom, of which the world partly makes use, some more than others, some of this and others of that. . . the First is the Common Philosophy of Aristotle, of Plato, and of our own time, which is

> but a Cagastrian Philosophy, Speculation, and Phantasy, with which, even at the present day, all the Schools are filled, and by which they are befouled, and beloved youth thereby led astray. The same is inane, erroneous, empty chatter; and far removed from the foundation of Truth. Even at the present day it is blasphemously defended, tooth and nail, with all sorts of opinions, ideas, imaginations, and erroneous thoughts of the old heathen (who were held to be Sages), which were accepted as the Truth. . .
>
> This Philosophy, although, from my youth up, it was earnestly and diligently inculcated, and forced upon me, in the Schools (as unfortunately occurs to others at the present day), yet, by special interposition of the Holy Spirit, it became so suspected by me that I never would, nor could, torture my head, mind, and soul with it, nor persuade my heart that the same was a sacred thing, nor cleave unto it as others did; but, according to my childish judgment, let the matter rest there until, about the year 1587 or 1588, another philosophy came into my hands. At the same time I had, in my own mind, firmly resolved not to remain the least among my fellow scholars, but in due time to graduate in advance of all.

This sounds very much like Bacon's famous letter to his uncle Lord Burghley, in which he says "I have taken all knowledge for my province." Figulus continues:

> But it has pleased God otherwise in His Divine Providence, and all sorts of impediments on the part of my superiors hindered the course of my studies, until at last, in 1587-88, the books and writings of Theophrastus, of Roger Bacon, and of M. Isaac the Hollander, fell into my hands; in which I, especially in medicine (for they wrote about the Universal Stone and Medicine), saw and found a better foundation, and yet understood it not at first. But I took such a liking to the subject that I resolved not to die, nor yet to take my ease, until I had obtained this Universal Stone and Blessed Heavenly Medicine. However, the poverty of my parents and the impossibility of obtaining the necessary funds (for at that time but few princes and nobles patronized this study) compelled me unwillingly to relinquish my plan, although I was so eager for it that, for many months, I could not sleep on account of it. At last, in 1590, I found myself plunged by the devil and his friends into great misery, misfortune, and sickness, out of which God mercifully helped me when my death would have been preferred to my recovery, and when, from reasons of poverty, I had been held to commerce against my will, by my relatives, suffering all manner of persecution, partly from the Anti-Christian mob, partly from false

brethren, wife and friends, tortured, plagued and agitated, and thus thoroughly tried by the devil.

But having been rescued from the same by God's fatherly care, I turned my attention for some years to poetry, whereunto, when I found that it was irksome to all, I said good bye...

Title page, *New Atlantis* "In time the hidden truth is revealed"
"Truth is the daughter of time, not authority"

Conclusion: The hero restored

> We are still far from knowing even half enough about Lord Bacon—the first realist in the grand sense—to know everything he did, everything he willed, and everything that went on in the depths of his soul… Let the critics go to hell!
>
> Freidrich Nietzsche, *Ecce Homo*

The Victorian marginalization of Bacon

The precipitous decline of Francis Bacon's reputation in the nineteenth century is one of the more curious episodes in modern intellectual history. For nearly two centuries after his death, Bacon was revered across Europe as a foundational figure of modern thought. Writers of the French Enlightenment spoke of him in superlatives: Voltaire called him "the father of the experimental philosophy," Denis Diderot declared that the

Encyclopédie "owe[s] most to the Chancellor Bacon," and d'Alembert praised him as "the greatest, the most universal, and the most eloquent of philosophers." By the late eighteenth century, Bacon was widely regarded not merely as an influence on science, but as one of the principal architects of modernity itself.

And yet, by the late nineteenth and twentieth centuries, Bacon's status had dramatically eroded. He came to be treated as a transitional figure—important, perhaps, but philosophically shallow, ethically compromised, and intellectually superseded. As one editor of the *Oxford Francis Bacon* has remarked, Bacon was "relegated to an intellectual *salon des refusés* from which he has been hard put to escape." How did such a reversal occur?

The turning point is almost universally acknowledged: Thomas Macaulay's long biographical essay on Bacon, published in 1837 in the *Edinburgh Review*. The essay's influence has been difficult to overstate. It established the tonal template, confident, prosecutorial, morally assured, through which Bacon would be read for generations. Although Macaulay conceded the brilliance of Bacon's intellect, calling it "the most exquisitely constructed intellect that has ever been bestowed on any of the children of men," he relentlessly portrayed Bacon's character as small, servile, and ethically suspect. The cumulative effect was devastating: Bacon's thought was preserved, but his authority was hollowed out.

What is too rarely emphasized is the context in which this judgment was delivered. Macaulay was not, at the time, a detached literary critic surveying the past from a quiet English study. He wrote the essay while serving in India, as a member of the Supreme Council of the colonial government, deeply embedded in the machinery of empire. Just two years earlier, he had authored the notorious *Minute on Indian Education* (1835), in which he dismissed the entire corpus of Sanskrit, Arabic, and Persian learning as inferior to "a single shelf of a good European library." This was not a casual provocation but a foundational text of British colonial ideology, justifying the imposition of English education and Western epistemic authority across the subcontinent.

The Bacon essay belongs to this same ideological moment, the dawn of the Victorian era, a time of rapid expansion of the empire. It was published in a journal explicitly committed to Whig progressivism, moral self-confidence, and a linear narrative of Western advancement. Its extraordinary length and rhetorical force mark it not as neutral criticism but as a moral indictment, conducted with the assurance of a man convinced that history itself stands behind him. Macaulay was not merely evaluating Bacon; he was consolidating a story about where modernity came from and what kinds of minds were permitted to stand at its origin.

Within that story, Bacon posed a problem.

Macaulay must be understood not merely as a historian, but as a principal architect of Victorian moral narrative. Writing at the moment when Britain was consolidating its imperial self-image, Macaulay inherited, and energetically reinforced, a conception of English history that depended upon the dignity, coherence, and moral legibility of its founding figures. In this context, Francis Bacon posed a particular problem. Bacon was too foundational to be ignored, yet too disruptive to be treated neutrally: a figure who stood at the origins of modern science, administrative rationality, and institutional knowledge, yet whose life unfolded in close proximity to court favoritism, dynastic anxiety, and political improvisation. Macaulay's assault on Bacon's character, remarkable for its vehemence and its moral absolutism, thus reads less as detached historical judgment than as an act of narrative containment, one that reduced Bacon's intellectual stature by converting him into a moral exemplum rather than a structural agent of modernity.

If Bacon were in any sense the concealed or displaced outcome of an illicit and politically dangerous union between Elizabeth I and Robert Dudley, the vehemence of such containment becomes intelligible at a deeper, archetypal level. The figure that would emerge is not merely a compromised courtier, but one uncannily close to the ancient heroic pattern of the exposed child or displaced heir: born under conditions that threaten sovereignty, separated from legitimate succession, and redirected into a compensatory vocation as lawgiver, culture-bringer, or founder of institutions rather than dynasties. Such figures, Oedipus, Moses, Romulus, are never comfortably assimilated into official historiography, for they expose the contingent, vulnerable origins of authority itself. To acknowledge even the possibility of Bacon occupying this structural position would be to admit that the Tudor settlement, like so many regimes before it, rested not on immaculate continuity but on suppression, substitution, and symbolic reconfiguration. Macaulay's Bacon, stripped of heroic ambiguity and reduced to personal failing, thus performs an essential Victorian function: it neutralizes a potentially archetypal threat by insisting that Bacon's fall was moral rather than structural, individual rather than dynastic, and thereby preserves the retrospective majesty of the monarchy at the cost of historical depth.

What Macaulay consistently minimizes, or actively distorts, are precisely those elements of Bacon's thought that resist incorporation into a triumphalist Whig narrative: his allegorical and mythic imagination in *The New Atlantis* and *The Wisdom of the Ancients*; his insistence that knowledge is provisional, symbolic, and morally constrained; his openness to non-Greek and non-Latin sources of wisdom; and his sympathy for the idea of a *prisca sapientia*, an ancient and widely diffused wisdom underlying multiple traditions. This Bacon is not a clean ancestor of nineteenth-century scientism. He is a mediator, a synthesist, and most dangerously, a moral legislator.

Such a figure sits uneasily with the ideological needs of empire. By the 1830s, British colonial authority rested increasingly on a sharp epistemic boundary: the West as rational, empirical, and progressive; the East as mystical, traditional, and backward. This dichotomy was politically indispensable, even as it was intellectually fragile. The emerging field of Indo-European philology had already begun to undermine it, revealing deep linguistic (and by implication cultural) continuities between Sanskrit, Greek, Latin, and the Germanic languages. As scholars such as Thomas McEvilley have shown, this discovery generated profound epistemic anxiety. If knowledge was cumulative and civilizational rather than uniquely Western, the moral rationale for empire became harder to sustain.

A Bacon who blurs the line between reason and myth, who insists that knowledge must be bounded by ethics and humility, and who refuses to identify progress with domination is therefore not merely inconvenient. He is destabilizing. Flattening Bacon into a rhetorically gifted but morally compromised proto-scientist neutralizes that threat. It preserves Bacon's utility while stripping him of authority.

And there is an additional layer, one that has never been adequately explored. Macaulay's essay appeared in 1837, the very year Queen Victoria ascended the throne, inaugurating a period of unprecedented imperial expansion. Within two years, Britain would fight the First Opium War, a conflict that laid bare the ethical contradictions of an empire claiming moral superiority while enforcing narcotics trade at gunpoint. The legitimacy of the crown, and of the civilizing mission, was therefore not an abstract matter. It was a live political concern.

Bacon and the conflict between science and religion

Francis Bacon offers a way of dissolving the modern conflict between science and religion, not by compromise, but by re-ordering their jurisdictions. The quarrel, as Bacon saw with unusual clarity, arises from a category error: from asking of Scripture what belongs to nature, and of nature what belongs to Scripture. When either domain is forced to answer questions it was never meant to address, both truth and piety are corrupted. Bacon's solution is not secularization, but discipline.

At the heart of his thought lies the distinction between God's two books: the Book of Scripture and the Book of Nature. Scripture is given for salvation, moral formation, and the knowledge of God's will; nature is given for inquiry, experiment, and the lawful expansion of human power. To confuse the two is dangerous in both directions. To read the Bible as a manual of physics invites superstition and intellectual paralysis; to read nature as a substitute for revelation invites pride and metaphysical overreach. Bacon insists that each book must be read according to its proper method, with humility before its limits.

This division does not diminish religion; it protects it. By refusing to anchor theology in outdated natural philosophies, Bacon shields faith from the inevitable revisions of scientific discovery. Truth about the natural world may change as instruments improve and methods refine, but divine truth does not thereby collapse, because it was never grounded in empirical description to begin with. In this sense Bacon anticipates the modern crisis of belief and pre-empts it: faith fails not because science advances, but because faith was mislocated.

Equally important, Bacon imposes a moral discipline on science itself. Knowledge of nature is not license, but stewardship. The goal of inquiry is not domination for its own sake, but the relief of the human estate. Power gained without moral orientation is not enlightenment but degeneration. Bacon repeatedly warns that the Fall was not merely a loss of innocence, but a loss of knowledge rightly ordered; the task of science is therefore restorative, not transgressive. It proceeds by obedience to nature's laws, not by metaphysical rebellion against God.

What Bacon offers, then, is neither Enlightenment triumphalism nor medieval synthesis, but a constitutional settlement. Science is freed from theological micromanagement, and theology is freed from scientific hostage-taking. Each gains autonomy without hostility. The result is not a cold, disenchanted world, but one in which inquiry can proceed vigorously without eroding reverence, and belief can endure without fear of discovery. Seen in this light, Bacon's project is deeply Christian, even when it is most radical. He does not abolish mystery; he relocates it. He does not deny transcendence; he refuses to counterfeit it. By restoring boundaries rather than collapsing them, Bacon allows modern science to emerge without requiring modern unbelief—and he allows religion to survive modernity without retreating into obscurantism.

There is an almost too-perfect irony at the level of symbol. In both Judaism and Islam, the pig is a creature placed outside the bounds of lawful consumption. It marks a line: between the clean and the unclean, the permitted and the forbidden, the ordered life and what must be excluded to preserve it. Francis Bacon, by sheer accident of name, stands on the wrong side of that line. Bacon is, quite literally, off limits.

Taken crudely, this would mean nothing. Taken symbolically, it means something precise. Dietary law is one of the oldest ways cultures encode obedience, boundary, and identity. What may be eaten is not a matter of nutrition but of covenantal discipline. To eat is to participate; to abstain is to remain within the law. In this sense, the forbidden animal becomes a sign of fidelity to inherited order. Bacon's intellectual project occupies a similar fault line. He does not reject divine law, but he refuses to confine truth to inherited textual authority alone. By insisting that nature itself is a legitimate object of disciplined inquiry, Bacon crosses a boundary that law-centered traditions have historically guarded with care. His appeal is not to commentary but to experiment, not to received interpretation but to fresh interrogation of creation. What is at stake is not impiety, but jurisdiction: who is authorized to know, and by what means. Seen this way, Bacon's "forbidden" status is not accidental, but emblematic. He represents a form of knowledge that cannot be regulated by purity codes, whether ritual, legal, or scholastic. His science cannot be kosher or halal, not because it is immoral, but because it answers to a different mode of authorization. It is knowledge without prior permission. That is precisely what makes it powerful—and threatening.

The deeper point is not about pork, nor about exclusion, but about the limits of law as a regulator of truth. Law excels at preserving identity, continuity, and obedience. It is less well suited to discovery. Bacon does not deny the necessity of law; he denies its sufficiency. Revelation, in his framework, governs salvation and morals. Nature governs works and powers. To confuse the two is to turn obedience into stagnation. In that sense, Bacon does not "refute" Judaism or Islam any more than he refutes medieval Christianity. He exposes a shared temptation across law-centered civilizations: the desire to protect truth by enclosing it. His wager is that God is not honored by enclosure, but by faithful exploration of what He has made. If some kinds of knowledge must remain symbolically "forbidden" to preserve a legal order, Bacon's reply is simple and unsettling: then the order, not the knowledge, must eventually change.

Why Bacon's story needs to be told

Modernity prefers a simple origin story. According to that story, Europe emerged from superstition into reason, from authority into method, from myth into science. Francis Bacon appears briefly in this narrative as a transitional figure: a gifted rhetorician who gestured toward empiricism before handing the project to more rigorous minds. He is acknowledged, then quietly set aside.

This book began from the conviction that this story is not merely incomplete, but distorting. Francis Bacon was not a naïve forerunner who happened to guess the direction of science. He was a lawgiver of modern knowledge: a thinker who grasped that the transformation of inquiry would reorganize institutions, ethics, power, and the human imagination itself. He did not simply ask how we might know more. He asked how knowledge is formed, corrupted, disciplined, transmitted, and governed—and what becomes of a civilization once knowledge is amplified into power.

Those questions are no longer antiquarian. They are our questions.

We live inside the world Bacon foresaw with unsettling clarity. Our systems of knowledge are collective rather than individual, technologically mediated, institutionally organized, economically incentivized, and politically consequential. Meaning is encoded into abstract differences, transmitted across distance, amplified by machinery, and converted into authority. Bacon understood this not metaphorically but structurally, long before electricity, computation, or digital networks.

His insight that any system capable of binary difference could carry meaning was not a parlor trick. It was a general theory of representation. Information theory did not begin with computers; it began as a philosophical insight into how mind enters matter. Yet Bacon paired this technical brilliance with an anxiety that modern culture has largely forgotten: that method alone is not wisdom, and that power untethered from moral orientation becomes domination. This double vision, methodological audacity joined to ethical restraint, is precisely what later narratives suppress.

Twentieth-century philosophy of science often treated Bacon as a foil. Karl Popper dismissed him as the emblem of naïve inductivism, a thinker who believed science proceeds by the simple accumulation of observations. Thomas Kuhn inherited a similar picture, relegating Bacon to the prehistory of science while locating genuine insight in paradigms, revolutions, and community-bound norms. Yet this Bacon never existed. Bacon explicitly rejected simple enumeration. He insisted on negative instances, staged inquiry, disciplined experiment, and the active reform of the mind. Observation, for Bacon, was never innocent; it required training, correction, and institutional support. His theory of the Idols (of the Tribe, the Cave, the Marketplace, and the Theatre) amounts to one of the earliest systematic psychologies of error. Knowledge fails not merely because of ignorance, but because of language, custom, authority, ambition, and collective intoxication with success.

Popper's emphasis on error and refutation, and Kuhn's emphasis on paradigm-bound perception, both operate inside a Baconian problem-space. What they strip away is Bacon's insistence that epistemology cannot be separated from ethics, institutions, and power. By narrowing Bacon into a crude inductivist, later philosophers made modern science appear more innocent than it is, and more self-justifying than Bacon ever believed.

Kuhn's paradigms perform, in historical form, the same work Bacon's Idols perform psychologically and institutionally. Paradigms shape what counts as evidence, which questions are thinkable, and which anomalies can be ignored. Scientific revolutions occur not because reason gradually accumulates facts, but because inherited frameworks collapse under pressure. Bacon knew this. He simply refused to treat it as a neutral process. For him, the overthrow of idols was a moral labor requiring humility, patience, and restraint. Kuhn describes revolutions; Bacon asks how they should be governed. This difference matters. Once Bacon is removed, scientific change appears as a sociological inevitability rather than a moral responsibility. Knowledge becomes something that happens to us, rather than something for which we are accountable.

If analytic philosophy minimized Bacon by caricature, critical theory condemned him by symbol. In the *The Dialectic of Enlightenment*, Max Horkheimer and Theodor Adorno cast Bacon as the progenitor of instrumental reason: the moment when knowledge becomes domination, nature becomes raw material, and reason collapses into technique. There is truth in this charge, but it is incomplete. Bacon did indeed link knowledge and power; he did not pretend otherwise. Yet Horkheimer and Adorno's critique depends on severing that link from Bacon's moral framework. They treat domination as the inevitable outcome of method, rather than as the result of method divorced from its governing ends.

Bacon anticipated their fear. His repeated insistence that knowledge be oriented toward the relief of the human estate, governed by humility and constrained by moral purpose, shows that he did not imagine technique as self-justifying. The Frankfurt School inherits Bacon's diagnosis of modern power while rejecting his attempt to discipline it. In doing so, they reproduce the very fatalism they seek to escape.

One reason Bacon remains difficult to classify is his refusal to abandon symbol. His allegories, fables, and mythic writings are often dismissed as vestigial medievalism. In fact, they reflect a sophisticated understanding of cognition. Humans do not think only in propositions; they think in images, narratives, and structures of meaning that exceed formal logic.

By flattening Bacon into a narrow empiricist, later positivism erased this dimension. Modernity did not begin when myth was abandoned; it began when myth was repressed. The result has been a technical culture rich in power and poor in orientation.

Bacon offers a way of dissolving the modern conflict between science and religion not by compromise, but by reordering their jurisdictions. Scripture governs salvation and moral formation; nature governs works and powers. Each must be read according to its proper method. To confuse them is to corrupt both. This is not secularization but discipline. By refusing to anchor theology in provisional natural philosophies, Bacon protects faith from scientific revision. By refusing to derive moral ends from technical success, he protects science from hubris. What Bacon proposes is a constitutional settlement rather than a culture war.

Bacon's most neglected contribution is his insistence that knowledge requires governance before it requires discovery. In *Valerius Terminus*, he likens knowledge to a powerful flood whose channels must be shaped before it is released. Ethics cannot be merely reactive; it must be architectural. Knowledge, for Bacon, is a trust rather than a possession. Its legitimacy depends on use, orientation, and public benefit. Science exists not for domination, prestige, or private gain, but for the relief of suffering and the common good. This vision stands in sharp contrast to both naïve instrumentalism and panicked regulation.

Early historians of the Royal Society understood this. They cast Bacon in Mosaic terms: not as one who enters the promised land, but as one who articulates the law under which others may proceed safely. This image captures Bacon's true position. He stands at the threshold between medieval metaphysics and modern science, between faith and experiment, between inherited authority and disciplined freedom. Later generations resolved this tension by choosing sides, secularizing Bacon into a proto-positivist or condemning him as an architect of domination. What has been lost is his insistence that knowledge and moral formation cannot be safely separated.

The modern world is saturated with the fruits of Baconian method, yet increasingly uncertain about its ends. We possess unprecedented powers to intervene in nature, the body, and the mind, while lacking a shared language for restraint or wisdom. This is precisely the imbalance Bacon feared. To recover Bacon is not to retreat from modernity, but to recover its conscience. He offers not a ready-made solution, but a model of responsibility at the origin of modern power. Knowledge is a vocation. Method is a moral discipline. Inquiry is a form of service. Bacon's legacy remains unresolved because the struggle he inaugurated is still underway. To understand him more fully is to understand ourselves more honestly and to reopen questions about knowledge, responsibility, and human purpose that we can no longer afford to ignore.

Appendix A: *Anti-Machiavel*

Parallelisms

Bacon, *Advancement of Learning*:

As for evil arts, if a man would set down for himself that principle of Machiavel, "That a man seek not to attain virtue itself, but the appearance only thereof; because the credit of virtue is a help, but the use of it is cumber"... or that other protestation of L. Catilina, to set on fire and trouble states, to the end to fish in droumy waters, and to unwrap their fortunes.

Anti-Machiavel:

As for peace, these people never like it, for they always fish in troubled water, gathering riches and heaps of the treasures of the realm while it is in trouble and confusion.

We should not then see France to be governed and ruled by strangers, as it is; we should not feel the calamities and troubles of civil wars and dissentions, which they enterprise to maintain their greatness and magnitude, and to fish in troubled water.

~

Bacon, *Advancement of Learning*:

Machiavel had reason to put the question, "which is the more ungrateful towards the well-deserving, the prince or the people?" though he accuses both of ingratitude. The thing does not proceed wholly from the ingratitude either of princes or people; but it is generally attended with the envy of the nobility; who secretly repine at the event, though happy and prosperous, because it was not procured by themselves .

Anti-Machiavel:

But I must say that sometimes such changes have been procured upon envy, rather than upon just complaint against those who governed; and such envies often proceed when kings govern themselves by men of base hand, as they call them, for then princes and great lords are jealous.

~

Bacon, "Of Seditions and Troubles":

Also, as Machiavel noteth well, when princes, that ought to be common parents, make themselves as a party, and lean to a side, it is as a boat that is overthrown by uneven weight on the one side... For when the authority of princes is made but an accessary to a cause, and that there be other bands that tie faster than the band of sovereignty, kings begin to be put almost out of possession.

Anti-Machiavel:

For if he nourishes partialities among his subjects, he cannot possibly carry himself so equally towards both parties, but in them both will be jealousy and suspicion. Each party will esteem the other to be more favored, whereupon he will hate his prince, and by that means it may come to pass that the prince shall be hated by both parties; and so both the one and the other shall machinate his ruin, which he can hardly shun, having all their evil wills.

~

Bacon, *Advancement of Learning*:

But that opinion I may condemn with like reason as Machiavel doth that other, that moneys were the sinews of wars; whereas (saith he) the true sinews of the wars are the sinews of men's arms, that is, a valiant, populous, and military nation; and he voucheth aptly the authority of Solon, who when Croesus shewed him his treasury of gold said to him, that if another came that had better iron he would be master of his gold.

Anti-Machiavel:

And although Machiavelli in a certain place where he speaks of war, maintains that the common saying is false, that money is the sinews of war; this hinders not, but what we say may be true.

The great treasures of king Croesus of Lydia incited him to war against king Cyrus of Persia and Media, to his own destruction.

~

Bacon, *Advancement of Learning*:

So in the fable that Achilles was brought up under Chiron the Centaur, who was part a man and part a beast: expounded ingeniously but corruptly by Machiavel, that it belongeth to the education and discipline of princes to know as well how to play the part of the lion in violence and the fox in guile, as of the man in virtue and justice.

Anti-Machiavel:

But should we call this beastliness or malice, what Machiavelli says of Chiron? Or has he read that Chiron was both a man and a beast? Who has told him that he was delivered to Achilles to teach him that goodly knowledge to be both a man and a beast?

~

Bacon, *Advancement of Learning*:

Concerning want, and that it is the case of learned men usually to begin with little and not to grow rich so fast as other men, by reason they convert not their labours chiefly to lucre and increase; it were good to leave the common place in commendation of poverty to some friar to handle, to whom much was attributed by Machiavel in this point, when he said, that "the kingdom of the clergy had been long before at an end, if the reputation and reverence towards the poverty of friars had not borne out the scandal of the superfluities and excesses of bishops and prelates."

Anti-Machiavel:

These mendicants then, being obliged and restrained unto poverty by a solemn vow which they made at their profession in their orders, they are so annexed, united, and incorporated in it and with it, that never after they could be never so little separated or dismembered, what diligence or labor soever they used to do it. Hereof they have found themselves much troubled and sorrowful, for howsoever gallant and goodly the *Theorique* of Poverty is, yet in practice they have found it a little too difficult and hard.

~

Bacon, *Advancement of Learning*:

And therefore the form of writing which of all others is fittest for this variable argument of negotiation and occasions is that which Machiavel chose wisely and aptly for government; namely, discourse upon histories or examples... And it hath much greater life for practice when the discourse attendeth upon the example, than when the example attendeth upon the discourse. For this is no point of order, as it seemeth at first, but of substance. For when the example is the ground, being set down in an history at large, it is set down with all circumstances, which may sometimes control the discourse thereupon made and sometimes supply it, as a very pattern for action; whereas the examples alleged for the discourse's sake are cited succinctly and without particularity, and carry a servile aspect toward the discourse which they are brought in to make good.

Anti-Machiavel:

Yet although the maxims and general rules of the political art may somewhat serve to know well to guide and govern a public estate, whether a principality or free city, yet they cannot be so certain as the maxims of the mathematicians, but are rules rather very dangerous, yea pernicious if men cannot make them serve and apply them unto affairs as they happen to come; and not to apply the affairs unto these maxims and rules. For the circumstances, dependencies, consequences, and antecedents of every affair and particular business, are all for the most part diverse and contrary; so that although two affairs be like, yet men must not therefore conduct and determine them by one same rule or maxim, because of the diversity and difference of accidents and circumstances.

~

Bacon, *Novum Organum*:

There are and can be only two ways of searching into and discovering truth. The one flies from the senses and particulars to the most general axioms, and from these principles, the truth of which it takes for settled and immovable, proceeds to judgment and middle axioms. And this way is now in fashion. The other derives axioms from the senses and particulars, rising by a gradual and unbroken ascent, so that it arrives at the most general axioms last of all.

Anti-Machiavel:

Aristotle and other philosophers teach us, and experience confirms, that there are two ways to come unto the knowledge of things. The one, when from the causes and maxims, men come to knowledge of the effects and consequences. The other, when contrary, by the effects and consequences we come to know the causes and maxims… The first of these ways is proper and peculiar unto the mathematicians, who teach the truth of their theorems and problems by their demonstrations drawn from maxims, which are common sentences allowed of themselves for true by the common sense and judgment of all men. The second way belongs to other sciences, as to natural philosophy, moral philosophy, physic, law, policy, and other sciences.

~

Bacon, "Of Discourse":

It is good, in discourse and speech of conversation, to vary and intermingle speech of the present occasion with arguments, tales with reasons, asking of questions with telling of opinions, and jest with earnest.

Anti-Machiavel:

For as Cato says, amongst serious things joyous and merry things would be sometimes mixed.

~

Bacon, *New Atlantis*:

The reverence of a man's self is, next religion, the chiefest bridle of all vices.

Anti-Machiavel:

Behold then the consequence of that most wicked and detestable doctrine of that wicked atheist; which is to bring all people to a spite and a mockery of God and his religion, and of all holy things, and to let go the bridle to all vices and villainies.

~

Bacon, *De augmentis scientiarum*:

Constancy is the foundation on which virtues rest.

Anti-Machiavel:

I will then presuppose that constancy is a quality which ordinarily accompanies all other virtues; it is, as it were, of their substance and nature.

~

Bacon, "Of Adversity":

Prosperity doth best discover vice, but adversity doth best discover virtue.

Anti-Machiavel:

Adversity also is a true touchstone to prove who are feigned or true friends, for when a man feels labyrinths of troubles fall on him, dissembling friends depart from him, and those who are good abide with him, as said the poet Euripides: Adversity the best and certain'st friends doth get, prosperity both good and evil alike doth fit.

~

Bacon, "Of Great Place":

It is much true which was anciently spoken: A place showeth the man, and it showeth some to the better, and some to the worse.

Anti-Machiavel:

And we see but too much by experience that the old proverb is true, honors change manners.

~

Bacon, "Of Suspicion":

But this would not be done to men of base natures; for they, if they find themselves once suspected, will never be true.

Anti-Machiavel:

For the best fortress that is, is not to be thought evil by subjects; and if a prince is once thought so, there is no fortress that can save him.

~

Bacon, *Apophthegms New and Old*:

Mr. Bettenham used to say, that riches were like muck: when it lay upon an heap, it gave a stench, and ill odour; but when it was spread upon the ground, then it was the cause of much fruit.

Bacon, "Of Riches":

Of great riches there is no real use, except it be in the distribution; the rest is but conceit.

Anti-Machiavel:

Briefly, it is neither good nor profitable for a prince to heap up great treasures and riches enclosed in one place. And what then? must a sovereign prince be poor? No, but contrary, he has need to be rich and very opulent, for otherwise he shall be feeble and weak, and cannot make head against his enemies; but his riches and treasures must be in the purses and houses of his subjects.

~

Bacon, "Of Riches":

Men leave their riches either to their kindred, or to the public; and moderate portions prosper best in both. A great state left to an heir, is as a lure to all the birds of prey round about to seize on him, if he be not the better stablished in years and judgment.

Anti-Machiavel:

For it is neither good nor profitable that a prince treasures up heaps of riches; for it serves for a bait to draw unto him enemies, or to engender quarrels and divisions after him; and we often see that princes' great treasures are causes of more evil than good.

~

Bacon, *Advancement of Learning*:

It is true, that taxes levied by public consent, less dispirit, and sink the minds of the subject, than those imposed in absolute governments.

Anti-Machiavel:

It is certain that a prince may well make war and impose taxes without the consent of his subjects, by an absolute power; but it is better for him to use his civil power, so should he be better obeyed.

~

Anti-Machiavel:

True charity is joined unto faith, pity, and all other virtues.

Bacon, *Advancement of Learning*:

But these be heathen and profane passages, having but a shadow of that divine state of mind which religion and the holy faith doth conduct men unto, by imprinting upon their souls Charity, which is excellently called the bond of Perfection, because it comprehendeth and fasteneth all virtues together.

Anti-Machiavel:

But I must say that the Christian religion has launched and entered far deeper into the doctrine of good manners than the pagans and philosophers have done. For proof hereof I will take the maxim of Plato, that we are not only born for ourselves, but that our birth is partly for our country, partly for our parents, and partly for our friends. Behold a goodly sentence we can say no other; but if we compare it with the doctrine of Christians, it will be found maimed and defective. For what mention does Plato make of the poor? Where and in what place of this notable sentence does he set them? He speaks not at all of them; briefly, he would have it that our charity should be first employed towards ourselves, which they have well marked and followed who say that a well ordered charity begins with himself. But this is far from the doctrine which Saint Paul teaches the Christians when he says that charity seeks not her own; and also that which Christ himself commands, to love our neighbor as ourselves. Secondly Plato places our love towards our country, thirdly our love towards our parents, and lastly our friends. And what becomes of the poor? Let them do as they can, for Plato's charity stretches not to them.

~

Bacon, Speech on taking his seat in Chancery:

I will promise regularly to pronounce my decree within few days after my hearing and to sign my decree at the least in the vacation after the pronouncing, for fresh justice is the sweetest, and to the end that there be no delay of justice, nor any other means-making or laboring, but the labour of the counsel at the bar.

Anti-Machiavel:

And as we see that the greed of wicked magistrates is cause of the length of law cases, because they desire that the parties who plead before them should serve their turn as a cow for milk, it follows that the poor people are pillaged and eaten to the bones by those horseleeches. Also contrary, when the magistrate hates greed, he will dispatch and hasten justice to parties, and not hold them long in law, neither pillage and spoil them; a thing bringing great comfort and help to the people.

~

Bacon, "Of Counsel":

The wisest princes need not think it any diminution to their greatness, or derogation to their sufficiency, to rely upon counsel. God himself is not without, but hath made it one of the great names of his blessed Son; The Counsellor. Salomon hath pronounced that "in counsel is stability."

Anti-Machiavel:

For a prince, however prudent he is, ought not so much to esteem his own wisdom as to despise the counsel of other wise men. Solomon despised them not, and Charles the Wise always conferred of his affairs with the wise men of his council.

~

Bacon, *Advancement of Learning*:

But this appeareth more manifestly, when kings themselves, or persons of authority under them, or other governors in commonwealths and popular estates, are endued with learning. For although he might be though partial to his own profession, that said "then should people and estates be happy, when either kings were philosophers, or philosophers kings"; yet so much is verified by experience, that under learned princes and governors there have ever been the best times.

Anti-Machiavel:

I am content to presuppose that it is certain that there cannot come a better and more profitable thing to a people than to have a prince wise of himself; therefore, said Plato, men may call it a happy commonwealth when either the prince can play the philosopher, or when a philosopher comes to reign there.

~

Bacon, *Advancement of Learning*:

For howsoever it hath been ordinary with politic men to extenuate and disable learned men by the names of *Pedantes;* yet in the records of time it appeareth in many particulars, that the governments of princes in minority (notwithstanding the infinite disadvantage of that kind of state) have nevertheless excelled the government of princes of mature age, even for that reason which they seek to traduce, which is, that by that occasion the state hath been in the hands of *Pedantes*. For so was the state of Rome for the first five years, which are so much magnified, during the minority of Nero, in the hands of Seneca, a *Pedanti*: so it was again for ten years space or more, during the minority of Gordianus the younger, with great applause and contentation in the hands of Misitheus, a *Pedanti*: so was it before that, in the minority of Alexander Severus, in like happiness, in hands not much unlike, by reason of the rule of the women, who were aided by the teachers and preceptor.

Anti-Machiavel:

This may yet be better showed by the examples of many princes who have been of small wisdom and virtue, and yet notwithstanding have ruled the commonwealth well by the good and wise counsel of prudent and loyal counsellors wherewith they were served; as did the emperor Gordian the Young, who was created emperor at eleven years of age. Many judged the empire to be fallen into a childish kingdom, and so into a weakness and a bad conduction; but it proved otherwise, for this young emperor Gordian espoused the daughter of a wise man called Misitheus, whom he made the high steward of his household, and governed himself by his counsel in all his affairs; so that the Roman Empire was well ruled so long as Misitheus lived... I will not here repeat the example of the emperor Alexander Severus, who came to the empire very young, and under whom the affairs of the commonwealth were so well governed, by the means of good counsellors, as above said.

~

Bacon, *Advancement of Learning*:

The writing of speculative men of active matter for the most part doth seem to men of experience, as Phormio's argument of the wars seemed to Hannibal, to be but dreams and dotage.

Anti-Machiavel:

Herein it falls out to Machiavelli as it did once to the philosopher Phormio; who one day reading in the Peripatetic school of Greece, and seeing arrive and enter there Hannibal of Carthage (who was brought thither by some of his friends, to hear the eloquence of the philosopher), he began to speak and dispute with much babbling of the laws of war and the duty of a good captain, before this most famous captain, who had forgotten more than ever that proud philosopher knew or had learned. When he had thus ended his lecture and goodly disputation, as Hannibal went from the auditory one of his friends who had brought him there asked what he thought of the philosopher's eloquence and gallant speech. He said, "Truly I have seen in my life many old dotards, but I never saw one so great as this Phormio."

~

Bacon, *Advancement of Learning*:

For Machiavel noteth wisely, how Fabius Maximus would have been temporizing still, according to his old bias, when the nature of war was altered and required hot pursuit.

Bacon, *Apophthegms New and Old*:

Fabius Maximus being resolved to draw the war in length, still waited upon Hannibal's progress, to curb him; and for that purpose, he encamped upon the high grounds. But Terentius his colleague fought with Hannibal, and was in great peril of overthrow. But then Fabius came down the high grounds, and got the day. Whereupon Hannibal said, That he did ever think, that that same cloud that hanged upon the hills, would at one time or other, give a tempest.

Anti-Machiavel:

Seeing this, the Roman Senate sent against Hannibal Fabius Maximus, who was not so forward (and it may be not so hardy) as Flaminius or Sempronius were; but he was more wise and careful, as he showed himself. On his arrival he did not set upon Hannibal, who desired no other thing, but began to coast him far off, seeking always advantageous places. And when

Hannibal approached him, then would he show him a countenance fully determined to fight, yet always seeking places of advantage. But Hannibal, who was not so rash as to join with his enemy to his own disadvantage, made a show to recoil and fly, to draw him after him. Fabius followed him, but upon coasts and hills, seeking always not the shortest way, but that way which was most for his advantage. Hannibal saw him always upon some hill or coast near him, as it were a cloud over his head; so that after Hannibal had many times essayed to draw Fabius into a place fit for himself, and where he might give battle for his own good, and yet could not thereunto draw him, said: "I see well now that the Romans also have gotten a Hannibal; and I fear that this cloud, which approaching us, still hovers upon those hills, will one of these mornings pour out some shower on our heads."

~

Bacon, "Of the True Greatness of Kingdoms and Estates":

A civil war, indeed, is like the heat of a fever; but a foreign war is like the heat of exercise, and serveth to keep the body in health.

Anti-Machiavel:

Therefore a foreign war seems not to be very damaging, but something necessary to occupy and exercise his subjects; but domestic and civil wars must be shunned and extinguished with all our power, for they are things against the right of nature, to make war against the people of their own country, as he that does it against his own entrails.

~

Bacon, "Of Unity in Religion":

But we may not take up the third sword, which is Mahomet's sword, or like unto it; that is, to propagate religion by wars or by sanguinary persecutions to force consciences.

Bacon, "Advertisement Touching a Holy War":

I was ever of opinion, that the Philosopher's Stone, and a Holy War, were but the *rendez-vous* of cracked brains.

Anti-Machiavel:

But here may arise a question, if it is lawful for a prince to make war for religion, and to constrain men to be of his religion. Hereupon to take the thing by reason, the resolution is very easy; for seeing that all religion consists in an approbation of certain points that concern the service of God, it is certain that such an approbation depends upon the persuasion which is given to men thereof. But the means to persuade a thing to any man is not to take weapons to beat him, nor to menace him, but to demonstrate to him by good reasons and allegations what may induce him to a persuasion.

~

Bacon, "Of the Vicissitude of Things":

Surely there is no better way to stop the rising of new sects and schisms, than to reform abuses; to compound the smaller differences; to proceed mildly, and not with sanguinary persecutions; and rather to take the principal authors by winning and advancing them, than to enrage them by violence and bitterness.

Anti-Machiavel:

It is then very much expedient, if a man means to gather fruit, and do good by his speech, to use gentle and civil talk and persuasions, especially if he has to do with a prince or great man, who will not be gained by rigor (or as they say, by high wrestling), but by mild and humble persuasions.

~

Anti-Machiavel:

For that cynical liberty of some philosophers, who knew not how to reprehend and show men's faults but by taunts and bitter biting speeches, are not to be approved; as did that fool Diogenes, who ridiculously and triflingly talked with king Alexander the Great as if he had spoken to some simple burgher of Athens. And Callisthenes, whom Alexander led with him in his voyage into Asia, to instruct him in good documents of wisdom; who indeed was so austere, hard, and biting in all his remonstrances and reasonings, that neither the king nor any others could take in good part anything he taught.

Bacon, "A Proposal for Amending the Laws of England":

Callisthenes, that followed Alexander's court, and was grown in some displeasure with him, because he could not well brook the Persian adoration; at a supper, which with the Grecians was ever a great part talk, was desired, because he was an eloquent man, to speak of some theme; which he did, and chose for his theme the praise of the Macedonian nation;

which though it were but a filling thing to praise men to their faces, yet he did it with such advantage of truth, and avoidance of flattery, and with such life, as the hearers were so ravished with it that they plucked the roses off from their garlands, and threw them upon him; as the manner of applause then was. Alexander was not pleased with it, and by way of discountenance said, It was easy to be a good orator in a pleasing theme: "But," saith he to Callisthenes, "turn your stile, and tell us now of our faults, that we may have the profit, and not you only the praise"; which he presently did with such a force, and so piquantly, that Alexander said, The goodness of this theme had made him eloquent before; but now it was the malice of his heart, that had inspired him.

~

Anti-Machiavel:

When Alexander the Great departed from Macedonia to go to the conquest of Asia, he had all the captains of his army appear before him, and distributed to them almost all the revenue of his kingdom, leaving himself almost nothing. One of the captains, named Perdicas, said to him: "What then will you keep for yourself?" "Even hope," answered Alexander.

Bacon, *Advancement of Learning*:

Lastly, weigh that quick and acute reply which he made when he gave so large gifts to his friends and servants, and was asked what he did reserve for himself, and he answered, "Hope."

~

Anti-Machiavel:

Hereof we read a very remarkable example above others in Alexander the Great, king of Macedon. When he departed from his country to pass into Asia, to make war upon that great dominator Darius, he had with him first in his love among others, Craterus and Hephaestion, two gentlemen, his best friends and servants. Yet they were far different from each other, for Craterus was of a hard and sharp wit, severe, stoic, and melancholic, who altogether gave himself unto affairs of counsel, and indeed was one of the king's chief counsellors. But Hephaestion was a young gentleman, well complexioned and conditioned in his manners and behavior, of a good and quick wit, yet free of all care but to content and please the king in his sports and pastimes. They called Craterus the king's friend, and Hephaestion the friend of Alexander, as one that gave himself to maintain the person of his prince in mirths and pastimes, which were good for the maintenance of his health.

Bacon, *Advancement of Learning*:

For matter of policy, weigh that significant division, so much in all ages embraced, that he made between his two friends Hephaestion and Craterus, when he said, "that the one loved Alexander, and the other loved the king"; describing the principal difference of princes' best servants, that some in affection love their person, and others in duty love their crown.

~

Bacon, "A Proposal for Amending the Laws of England":

For the laws of Lycurgus, Solon, Minos, and others of ancient time, they are not the worse because grammar scholars speak of them.

Anti-Machiavel:

So is there great need of some Lycurgus or Solon to make those laws, men's wits are so wild, and their spirits so marvelously plentiful and fertile to bring forth contentions and differences, and so easily to dissent from each other.

~

Bacon, "Of Anger":

Anger is a kind of baseness, as it appears well in the weakness of whose subjects in whom it reigns.

Anti-Machiavel:

This vice of cruelty, proceeding from the weakness of those who cannot command their choler and passions of vengeance, and suffer themselves to be governed by them, never happened in a generous and valiant heart, but rather always in cowardly and fearful hearts.

~

Bacon, "Of Revenge":

Revenge is a kind of wild justice, which the more a man's nature runs to, the more ought law to weed it out.

Anti-Machiavel:

And if it were lawful for everyone to use vengeance, that would be to introduce a confusion and disorder into the commonwealth, and to enterprise upon the right which belongs to the magistrate, unto whom God has given the sword, to do right to everyone and to punish those who are faulty, according to their merits.

~

Bacon, "Of Revenge":

Public revenges are for the most part fortunate; as that for the death of Caesar; for the death of Pertinax; for the death of Henry the Third of France; and many more.

Anti-Machiavel:

Moreover, he exercised part of his cruelties in the revenge of the good emperor Pertinax, which was a lawful cause; yet withal he had in himself many goodly and laudable virtues, as we have in other places rehearsed.

~

Bacon, *History of the Reign of King Henry VII*:

After that Richard, the third of that name, king in fact only, but tyrant both in title and regiment, and so commonly termed and reputed in all times since, was by the Divine Revenge, favouring the design of an exiled man, overthrown and slain at Bosworth Field; there succeeded in the kingdom the Earl of Richmond, thenceforth styled Henry the Seventh.

Anti-Machiavel:

A similar punishment happened by the judgment of God to that cruel king Richard of England, brother of Edward IV... Yet that king, who despaired otherwise to be maintained in his estate, gave battle to the earl and was slain fighting, after he had reigned about a year. And the earl of Richmond went right to London with his victory, and the slaying of that tyrant; then he took out of the monastery Edward's two daughters, espoused the elder, and was straight made king of England, called Henry VII, grandfather of the most illustrious queen Elizabeth presently reigning.

~

Bacon, "Of Friendship":

The like or more was between Septimus Severus and Plautianus. For he forced his eldest son to marry the daughter of Plautianus; and would often maintain Plautianus in doing affronts to his son; and did write also in a letter to the senate, by these words: "I love the man so well, as I wish he may over-live me."

Anti-Machiavel:

The emperor Severus advanced Plautianus so high, that being great master of his household, the people thought he was the emperor himself, and that Severus was but his great master.

~

Bacon, "Of Friendship":

Augustus raised Agrippa (though of mean birth) to that height, as when he consulted with Maecenas about the marriage of his daughter Julia, Maecenas took the liberty to tell him, that "he must either marry his daughter to Agrippa, or take away his life: there was no third way, he had made him so great."

Anti-Machiavel:

And here that manner of electing friends which Augustus Caesar observed is worthy of observation. For he did not easily retain every man in his friendship and familiarity, but took time to prove and find their virtues, fidelity, and loyalty. Those he knew to be virtuous people, and who would freely tell him the truth of all things (as did that good and wise Maecenas), and who would not flatter him, but would employ good will sincerely in the charges he gave them—after he had well proved them, then would he acknowledge them his friends.

~

Bacon, *De augmentis scientiarum*:

When the prince is one who lends an easy and credulous ear without discernment to whisperers and informers, there breathes as it were from the king himself a pestilent air, which corrupts and infects all his servants. Some probe the fears and jealousies of the prince, and increase them with false tales.

Anti-Machiavel:

A marmoset, according to the language of our elders, is as much to say a reporter, murmurer, whisperer of tales behind one's back in princes' and great men's ears, which are false, or else not to be reiterated or reported.

~

Bacon, "Of the True Greatness of Kingdoms and Estates":

And, certainly those degenerate arts and shifts, whereby many counsellors and governors gain both favour with their masters and estimation with the vulgar, deserve no better name than fiddling; being things rather pleasing for the time, and graceful to themselves only, than tending to the weal and advancement of the state which they serve.

Anti-Machiavel:

First, there are those our ancient Frenchmen called janglers, which signifies as much as a scoffer, a trifler, a man full of words, or as we call them, long tongues, who by their jangling and babbling in rhyme or in prose give themselves to please great men, in praising and exalting them exceedingly, and rather for their vices than for their virtues.

~

Bacon, "Of Friendship":

So as there is as much difference between the counsel that a friend giveth, and that a man giveth himself, as there is between the counsel of a friend and of a flatterer; for there is no such flatterer as is a man's self, and there is no such remedy against flattery of a man's self as the liberty of a friend.

Anti-Machiavel:

And above all, men ought well to engrave in princes' minds that notable answer that Phocion made unto the king Antipater, who had required something of him which was not reasonable. "I would, sir, do for you service all that is possible for me, but you cannot have me both for a friend and a flatterer." As if he would say that they be two things far different, to be a friend and to be a flatterer, as in truth they are.

~

Bacon, *Ornamenta Rationalia*:

The coward calls himself a cautious man; and the miser says, he is frugal.

Anti-Machiavel:

And it helps to this persuasion that the flatterer always takes for the subject of his praises those vices which are in alliance and neighborhood with their virtues. For if the prince is cruel and violent, he will persuade him that he is magnanimous and generous, and such a one as will not put up with an injury. If the prince is prodigal, he will make him believe that he is liberal and magnificent, that he maintains an estate truly royal, and one that well recompenses his servants. If the prince is overgone in lubricities and lusts, he will say he is of a humane and manly nature, of a jovial and merry complexion, and of no saturnine complexion or condition. If the prince is covetous and an eater of his subjects, he will say he is worthy to be a great prince as he is, because he knows well how to make himself well obeyed. Briefly, the flatterer adorns his language in such sort that he will always praise the prince's vice by the resemblance of some virtue near thereunto. For most vices have a likeness with some virtue.

~

Bacon, *Ornamenta Rationalia*:

He that injures one, threatens an hundred… he of whom many are afraid, ought himself to fear many.

Anti-Machiavel:

Moreover, cruelty is always hated by everyone; for although it be not practiced upon all individuals, but upon some only, yet those upon whom it is not exercised cease not to fear when they see it executed upon their parents, friends, allies, and neighbors. But the fear of pain and punishment engenders hatred; for one can never love that whereof he fears to receive evil, and especially when there is a fear of life, loss of goods, and honors, which are the things we hold most precious.

~

Bacon, *Ornamenta Rationalia*:

He conquers twice, who restrains himself in victory.

Anti-Machiavel:

The clemency of a prince is the cause of the increase of his domination. Hereupon we read a memorable history of Romulus, who was so clement, soft, and gentle towards the people he vanquished and subjugated, that not only many individuals but the whole multitude of people submitted themselves voluntarily and unconstrainedly under his obedience. The same virtue was also the cause that Julius Caesar vanquished the Gauls; for he was so soft and gracious to them, and so easy to pardon, and used them every way so well, far from oppression, that many of that nation voluntarily joined themselves unto him, and by them he vanquished the others. When Alexander the Great made great conquests in Asia, most commonly the citizens of all great cities met him to present him with the keys of the towns; for he dealt with them in such clemency and kindness, without in any way altering their estates, that they liked better to be his than their own.

~

Bacon, *Advancement of Learning*:

When Periander, being consulted how to preserve a tyranny newly usurped, bid the messenger report what he saw; and going into the garden, cropped all the tallest flowers; he thus used as strong an hieroglyphic as if he had drawn it upon paper.

Anti-Machiavel:

Periander, having tyrannously obtained the crown of Corinth where he had no right, fearing some conspiracy against him, sent a messenger to ask advice of his great friend Thrasibulus, so to be assured master and lord of Corinth. Thrasibulus made him no answer by mouth; but commanding the messenger to follow him, he went into a field full of ripe corn, and taking the highest and most eminent ears there, he bruised them between his hands and wished the messenger to return to Periander, saying no more unto him. As soon as Periander heard of bruising the most ancient ears of corn, he presently conceived the meaning thereof; to wit, to overthrow and remove all the great men of Corinth who suffered any loss and were grieved at the change of the state; as indeed he did.

~

Bacon, *Advancement of Learning*:

And the virtue of this prince, continued with that of his predecessor, made the name of Antoninus so sacred in the world, that though it were extremely dishonoured in Commodus, Caracalla, and Heliogabalus, who all bare the name, yet when Alexander Severus refused the name because he was a stranger to the family, the Senate with one acclamation said, *Quomodo Augustus, sic et Antoninus*: in such renown and veneration was the name of these two princes in those days, that they would have had it as a perpetual addition in all the emperors' style.

Anti-Machiavel:

The very name of Antoninus was also so reverenced and loved by all the world, from father to son in generations after him many successive emperors caused themselves to be called Antonys, that rather they might be beloved of the people, though that name did not belong to them, nor were of the race or family of Antoninus; as did Diodumenus, Macrinus his son and companion in the empire, and as also did Bassianus and Geta, Severus' children, and Heliogabalus, they were all surnamed Antoninus. But as this name appertained not to them, so they held nothing of the virtues of that good emperor, with whose name they decked themselves.

Antony & Cleopatra:

Sir, sometimes when he is not Antony
He comes too short of that great property
Which still should go with Antony

~

Bacon, "Charge against Somerset":

So it appeareth likewise in Scripture, that the murder of Abner by Joab, though it were by David respited in respect of great services past, or reason of state, yet it was not forgotten.

Anti-Machiavel:

For the last example of this matter, I will set down that of Joab, David's nephew and constable, unto whom he did great services. Yet David commanded his son Solomon that he should put to death his cousin Joab, because of his perfidy.

~

Bacon, *Advancement of Learning*:

So likewise in the person of Solomon the king, we see the gift or endowment of wisdom and learning, both in Solomon's petition and in God's assent thereunto, preferred before all other terrene and temporal felicity. By virtue of which grant or donative of God Solomon became enabled not only to write those excellent parables or aphorisms concerning divine and moral philosophy, but also to compile a natural history of all verdure, from the cedar upon the mountain to the moss upon the wall (which is but a rudiment between putrefaction and an herb), and also of all things that breathe or move. Nay, the same Solomon the king, although he excelled in the glory of treasure and magnificent buildings, of shipping and navigation, of service and attendance, of fame and renown, and the like, yet he maketh no claim to any of those glories, but only to the glory of inquisition of truth; for so he saith expressly, "The glory of God is to conceal a thing, but the glory of the king is to find it out"; as if, according to the innocent play of children, the Divine Majesty took delight to hide His works, to the end to have them found out; and as if kings could not obtain a greater honour than to be God's playfellows in that game; considering the great commandment of wits and means, whereby nothing needeth to be hidden from them.

Anti-Machiavel:

Solomon was a king most wise, and a great philosopher; for he asked wisdom from God, who gave it in such abundance that besides being ignorant of nothing a prince should know to govern his subjects well, he also knew the natures of plants and living creatures, and was so cunning in all kinds of philosophy that his knowledge was admired through all the world. His prudence and wisdom made him so respected by all the great kings, his neighbors, that they esteemed themselves happy to do him pleasure and have his amity. By this means he maintained his kingdom in so high and happy a peace that in his time his subjects made no more account of silver than of stones, they had such store. And as for himself, he held so magnificent an estate, that we read not of any king or emperor that did the like.

~

Bacon, *Advancement of Learning*:

Dramatic poesy, which has the theatre for its world, would be of excellent use if well directed. For the stage is capable of no small influence both of discipline and of corruption. Now of corruptions in this kind we have enough; but the discipline has in our times been plainly neglected. And though in modern states play-acting is esteemed but as a toy, except when it is too satirical and biting; yet among the ancients it was used as a means of educating men's minds to virtue.

Anti-Machiavel:

After Solon had seen Thespis' first edition and action of a tragedy, and meeting with him before the play, he asked if he was not ashamed to publish such feigned fables under so noble, yet a counterfeit personage. Thespis answered that it was no disgrace upon a stage, merrily and in sport, to say and do anything. Then Solon, striking hard upon the earth with his staff, replied thus: "Yea but shortly, we that now like and embrace this play, shall find it practiced in our contracts and common affairs." This man of deep understanding saw that public discipline and reformation of manners, attempted once in sport and jest, would soon quail; and corruption, at the beginning passing in play, would fall and end in earnest.

~

Bacon, *Advancement of Learning*:

So again we find that many of the ancient bishops and fathers of the Church were excellently read and studied in all the learning of the heathen... it was the Christian Church, which amidst the inundations of the Scythians on the one side from the north-west, and the Saracens from the east, did preserve in the sacred lap and bosom thereof the precious relics even of heathen learning, which otherwise had been extinguished as if no such thing had ever been.

Anti-Machiavel:

But now I am desirous to know of this atheist Machiavelli, what was the cause that so many good books of the pagan authors were lost since the time of the ancient doctors of our Christian religion? Was it not by the Goths, who were pagans? For at their so many interruptions and breaking out of their countries, upon Gaul, Italy, and Spain, they wasted and burned as many books as they could find, being enemies of all learning and letters. And who within this hundred years has restored good letters contained in the books of the ancient pagans, Greeks, and Latins? Has it been the Turk, who is a pagan? It is well enough known that he is an enemy of letters, and desires none. Nay contrary, it has been the Christians who have restored them, and established them in the brightness and light wherein we see them today.

~

Bacon, "Of the Colours of Good and Evil":

So the Epicures say of the Stoics' felicity placed in virtue; that it is like the felicity of a player, who if he were left of his auditory and their applause, he would straight be out of heart and countenance; and therefore they call virtue *bonum theatrale* [public good].

Anti-Machiavel:

Briefly, a man may see within man an admirable and well ordained disposition of all the parts, and it brings us necessarily (whether we will or no) to acknowledge that there must be a God, a sovereign architect, who has made this excellent building; and by these considerations of natural things, whereof I do but lightly touch the points, the ancient philosophers, as the Platonists, Aristotelians, Stoics, and others, have been brought to the knowledge of a God and of his providence. And of all the sects of philosophers, there never was any which agreed not hereunto, unless the sect of the Epicureans, who were gluttons, drunkards, and whoremongers; who constituted their sovereign felicity in carnal pleasures, wherein they wallowed like brute beasts.

~

Bacon, "Of Custom and Education":

And therefore, as Machiavel well noteth, though in an ill-favoured instance, there is no trusting to the force of nature, nor to the bravery of words, except it be corroborate by custom. His instance is, that for the achieving of a desperate conspiracy, a man should not rest upon the fierceness of any man's nature, or his resolute undertakings, but take such a one as both had his hands formerly in blood.

Anti-Machiavel:

Catiline, a man devoid of all virtue and a bundle of all vice, resolving in his brain to be an exceedingly great man or altogether nothing, devised a conspiracy against his country and drew to his league many Roman gentlemen such as himself. Considering that he could not bring to effect his conjuration without declaring and communicating it to the chieftains of his aid, yet fearing that some of them would disclose it, he made them all take a most execrable oath, that thereby might be foreclosed from them all hope of retiring from his side. So he mixed wine with human blood in pots and made all his companions drink of it, and made them swear with an execration that they would never disclose the enterprise, but employ themselves with all their power to execute it. His partners, already culpable of human blood, were so secret that nothing would have been discovered if God had not permitted a harlot called Fulvia to draw certain words out of a conspirator's mouth, as she demanded of him where he lay the preceding nights. Being drunk, to enjoy his courtesan he disclosed to her that he had been in a company with whom he made an enterprise that would make him rich forever. As soon as Fulvia knew all the conjuration she disclosed it to the consul Cicero. Cicero did what he could to open all the enterprise, but the conspirators held so well their horrible oath that not one of so great a number would ever reveal a word. But yet Cicero found means to know all, by the declaration which the Allobroges made, who Catiline had appointed to furnish him with people for the execution. But the end of Catiline was such that he was slain fighting with a great number of others, and most of his accomplices were executed by justice. Briefly, all who have practiced that wicked doctrine of Machiavelli, to commit outrageous acts to be irreconcilable, their ends and lives have proved very tragedies.

~

Bacon, "Of the Colours of Good and Evil":

The ill that a man brings on himself by his own fault is greater; that which is brought on him from without is less. The reason is because the sting and remorse of the mind accusing itself, doubleth all adversity... So the poets in tragedies do make the most passionate lamentations, questioning, and torturing of a man's self... where the evil is derived from a man's own fault, there all strikes deadly inwards, and suffocateth.

Anti-Machiavel:

Men may see how an evil conscience leaves a man never in quiet. This wicked man, knowing that by his cruelty he had procured the hatred of his subjects, the wrath of God, and the enmity of all the world, was tormented in his conscience as of an infernal fury, which ever after fretted his languishing soul in the poor infected and wasted body.

~

Bacon, "Notes on the Present State of Christendom":

The division in his country [France] for matters of religion and state, through miscontentment of the nobility to see strangers advanced to the greatest charges of the realm, the offices of justice sold, the treasury wasted, the people polled, the country destroyed, hath bred great trouble, and like to see more.

Anti-Machiavel:

Besides the examples we read in histories, we know it by experience, seeing at this day all France fashioned after the manners, conditions, and vices of foreigners that govern it, and who have the principal charges and estates.

~

Advancement of Learning:

And as Alexander Borgia was wont to say, of the expedition of the French for Naples, that they came with chalk in their hands to mark up their lodgings not with weapons to fight ; so we like better, that entry of truth, which comes peaceably where the Minds of men, capable to lodge so great a guest, are signed, as it were, with chalk; than that which comes with Pugnacity, and forceth itself a way by contentions and controversies.

Anti-Machiavel:

King Charles VIII, in the voyage of Naples, which he made in his own person, conquered the realm of Naples almost without striking a stroke; and was received by all the people, and most of the nobility, as a messiah sent from God to deliver them from the cruel and barbarous tyranny wherein they had long endured under their kings, Alfonso and Ferdinand of Aragon, usurpers of that kingdom from the house of Anjou, to which Charles succeeded.

~

Bacon, *Novum Organum*:

We cannot command nature except by obeying her.

Anti-Machiavel:

It is evident enough that the felicity of a state lies wholly in well commanding and well obeying, whereupon results a harmony and concordance so melodious and excellent, that he who commands and he who obeys both receive contentment, pleasure, and utility. But to obey well depends wholly on well commanding, and cannot be without it; so commanding well depends on the prudence and wisdom of him that commands.

~

Anti-Machiavel:

And we see but too much by experience that the old proverb is true, honors change manners. You may see how the most gracious and courteous in the world, the most affable and officious to everyone (that is possible) while they are in base degree, after they are mounted into some high degree of honor and dignity they become rough and haughty, so much that those to whom they showed themselves facile and serviceable, they now seem not to know them, who before were their private friends and familiars. Such people have no good souls, but deserve that their fierceness and pride should dispossess them of that place unto which most commonly their dissembled humility and courtesy has advanced them. This vice is reprehensible, not only in a prince's officers but also in the prince himself, who ought not to put pride and fierceness upon that head whereupon the crown and diadem stands. For this the king Agamemnon is taxed and reprehended by Menelaus his brother, in a tragedy of Euripides, where he says thus:

> Most humble was thou in times past, and kissed each man's hand,
> Most humane, gentle, affable, to none thy gates did stand
> Shut up, to highest honor thou by such means sought to rise:
> But now thou honor has supreme, why proves thou so unwise,
> Another man straight to become, and change thy manners all?
> Yea human duty even to friends, by thee doth not befall.
> To good men that esteem good fame, this is not covenable,
> Chameleon like thy manners changed, thou to be so mutable.

Julius Caesar:

> He would be crowned.
> How that might change his nature, there's the question.
> It is the bright day that brings forth the adder,

And that craves wary walking. Crown him that,
And then I grant we put a sting in him
That at his will he may do danger with.
Th' abuse of greatness is when it disjoins
Remorse from power. And to speak truth of Caesar,
I have not known when his affections swayed
More than his reason. But 'tis a common proof
That lowliness is young ambition's ladder,
Whereto the climber-upward turns his face;
But when he once attains the upmost round
He then unto the ladder turns his back,
Looks in the clouds, scorning the base degrees
By which he did ascend. So Caesar may.

VINDICIAE,
CONTRA TY-
RANNOS:

SIVE,

*DE PRINCIPIS IN
Populum, Populique in Princi-
pem, legitima potestate,*

STEPHANO IVNIO
Bruto Celta, Auctore.

EDIMBVRGI, AN-
NO M. D. LXXIX.

Appendix B: *Vindiciae contra tyrannos*

The Huguenot tract on the right of resistance, *Vindiciae contra tyrannos* (translated as *A defense of liberty against tyrants*, 1579), was published in Basel with a false imprint of Edinburgh, under the pseudonym Stephanus Brutus Junius—alluding to both Marcus Junius Brutus (of *Julius Caesar*) and Lucius Junius Brutus, who deposed Tarquin and established the Roman Republic (*The Rape of Lucrece*). Machiavelli advised that "whoever takes up a tyranny and does not kill Brutus, and whoever makes a free state and does not kill the sons of Brutus, maintains himself for little time." The *Vindiciae*'s account of Tarquin reads:

> Tarquinius Superbus was therefore esteemed a tyrant, because being chosen neither by the people nor the senate, he intruded himself into the kingdom only by force and usurpation . . . The true causes why Tarquinius was deposed, were because he altered the custom, whereby the king was obliged to advise with the senate on all weighty affairs; that he made war and peace according to his own fancy; that he treated confederacies without demanding counsel and consent from the people or senate; that he violated the laws whereof he was made guardian; briefly that he made no reckoning to observe the contracts agreed between the former kings, and the nobility and people of Rome.

Anti-Machiavel:

> Tarquin, who enterprised to slay his father-in-law king Servius Tullius to obtain the kingdom of Rome, showed well by that act and many others that he was a very tyrant. . . when he changed his just and royal domination into a tyrannical government, he became a contemner and despiser of all his subjects, both plebian and patrician. He brought a confusion and a corruption into justice; he took a greater number of servants into his guard than his predecessors had; he took away the authority from the Senate; moreover, he dispatched criminal and civil cases after his fancy, and not according to right; he cruelly punished those who complained of that change of estate as conspirators against him; he caused many great and notable persons to die secretly without any form of justice; he imposed tributes upon the people against the ancient form, to the impoverishment and oppression of some more than others; he had spies to discover what was said of him, and punished rigorously those who blamed either him or his government.

The introduction to *The Rape of Lucrece* echoes these passages, and may reflect what T.S. Eliot called Shakespeare's "shameless lifting" from *Anti-Machiavel*:

> Tarquinius, for his excessive pride surnamed Superbus, after he had caused his own father-in-law Servius Tullius to be cruelly murdered, and, contrary to the Roman laws and customs, not requiring or staying for the people's suffrages, had possessed himself of the kingdom . . . the people were so moved, that with one consent and a general acclamation the Tarquins were all exiled, and the state government changed from kings to consuls.

The *Vindiciae*'s preface, which has been ascribed to the author of *Anti-Machiavel*,[20] includes an edict of Theodosius II and Valentinian III, whereby emperors became subject to Roman law; the edict is also transcribed in full in *Anti-Machiavel*. The *Vindiciae*'s preface challenged "the Machiavellians are free to descend into the arena: let them come forth. As we have said, we shall use the true and legitimate weapons of Holy Scripture"[21] *Anti-Machiavel*, on the other hand, "must fight against their impiety . . . not by assailing them with the arms of the holy Scripture . . . but by their proper arms and weapons" (that is, pagan authors). *Anti-Machiavel* and the *Vindiciae* draw from many of the same sources, biblical and classical; this in itself is unsurprising, but the similarities are so extensive as to indicate at the least a strong influence.

The *Vindiciae*'s authorship is still unresolved.[22] It was first attributed to François Hotman, author of the *Francogallia* (1573), another Huguenot "Monarchomach" treatise. Hotman's son Jean had been a tutor in the household of English ambassador Sir Amias Paulet, while Francis Bacon happened to be living there. Theodore Beza, author of *De jure magistratuum* (*The Right of Magistrates*, 1574), was then thought responsible; his connections with the Bacon family have been noted. The next candidate was Philippe du Plessis Mornay, a Huguenot author and diplomat who fled to England after the St. Bartholomew's Day massacres. During the peace negotiations at Poitiers in late 1577, Bacon met Mornay, who later invited Anthony Bacon to Montauban, and the two became good friends.[23] Finally Hubert Languet, or a collaboration between Languet and Mornay, was credited with the *Vindiciae*. Languet corresponded extensively with Sir Philip Sidney, a friend of Bacon's. Bacon himself has not been proposed as a possible author of the *Vindiciae*, but it is interesting to note that he had connections to all candidates, a fact that has so far been overlooked.

---

[20] By Mastellone (1969); see Victoria Kahn, "Reading Machiavelli: Innocent Gentillet's Discourse on Method." *Political Theory* 22, no. 4 (1994): 539-60.

[21] *Vindiciae, contra tyrannos*, tr. George Garnett, p. 11. Cambridge: Cambridge University Press, 1994. Other citations are from the 1648 English translation dubiously attributed to William Walker, supposed executioner of Charles I.

[22] Barker, Ernest. "The Authorship of the *Vindiciae Contra Tyrannos*." *Cambridge Historical Journal* 3, no. 2 (1930): 164-81. Also Garnett, ibid. pp. lv—lxxvi.

[23] See Daphne du Maurier, *Golden Lads* (1975).

Parallelisms

Vindiciae contra tyrannos:

Notwithstanding, the Machiavellians are free to descend into the arena: let them come forth. As we have said, we shall use the true and legitimate weapons of Holy Scripture, of the philosophy of ethics and of the laws of the commonwealth, of customs of nations, and of historical examples; then we shall boldly join battle with them on foot.

Anti-Machiavel:

I see well it is to no purpose to cite reasons against this atheist and his disciples, who believe neither God nor religion; wherefore, before I pass any further, I must fight against their impiety, and make it appear to their eyes, if they have any, not by assailing them with the arms of the holy Scripture—for they do not merit to be so assailed, and I fear to pollute the holy Scriptures among people so profane and defiled with impiety—but by their proper arms and weapons, whereby their ignorance and beastliness defends their renewed atheism.

~

Vindiciae contra tyrannos:

As for the characteristics of the method of teaching (I address myself to philosophers and disputants): from the effects and consequences he inferred the causes and major propositions or rules, in order to demonstrate the matter more clearly and definitively. He rendered it visible and comprehensible, as if ascending through certain degrees to the peak: so that in the manner of geometricians—whom he seems to have wanted to imitate in this matter—from a point he draws a line, from the line a plane, and from the plane he constitutes a solid.

Anti-Machiavel:

Aristotle and other philosophers teach us, and experience confirms, that there are two ways to come unto the knowledge of things. The one, when from the causes and maxims, men come to knowledge of the effects and consequences. The other, when contrary, by the effects and consequences we come to know the causes and maxims… The first of these ways is proper and peculiar unto the mathematicians, who teach the truth of their theorems and problems by their demonstrations drawn from maxims, which are common sentences allowed of themselves for true by the common sense and judgment of all men.

~

Vindiciae contra tyrannos:

In treating these questions we will bear in mind this old and, to be sure, perfect image of the governance of kingdoms, as a legitimate, chaste, and blameless matron without any excessive adornment; in its place these Machiavellians do not hesitate to present us with an illegitimate, painted, lewd, and wanton harlot. This ancient method of administering provinces, kingdoms, and empires was that of your ancestors; and princes who were well endowed with every sort of royal virtue carefully kept to it for as long as they lived, as something passed on from hand to hand.

Anti-Machiavel:

And we need not be abashed if those of Machiavelli's nation, who hold the principal estates in the government of France, have forsaken the ancient manner of our French ancestors' government, to bring France into use with a new form of managing and ruling their country, taught by Machiavelli.

~

Vindiciae contra tyrannos:

And clearly, in order that this majesty of the king and the ancient rights of the peoples should be restored in their entirety amongst the Gauls, some of your own compatriots have, as generals, led armies against that nation which, despising both God and man and buoyed up by the strengths and artifices of cunning and perfidy, wholly concentrated its talent, power, and force on reducing the Gauls—who are free by nature and entirely autonomous in their way of life and the laws and practices of antiquity—to a servitude of barbarous cruelty.

Anti-Machiavel:

The French were reputed to be frank and liberal, far from all servitude; but now our stupidity, carelessness, and cowardice make us servants and slaves to the most dastardly and cowardly nation of Christendom... Let us then stir up in ourselves the generosity and virtue of our valiant great grandfathers, and show that we are come from the race of those good and noble Frenchmen, our ancestors, who in time past have brought under their subjection so many foreign nations, and who so many times have vanquished the Italian race, who would make us now serve.

~

Vindiciae contra tyrannos:

A tyrant subverts the state, pillages the people, lays stratagems to entrap their lives, breaks promise with all, scoffs at the sacred obligations of a solemn oath, and therefore is he so much more vile than the vilest of usual malefactors… Therefore as Bartolus says, "He may either be deposed by those who are lords in sovereignty over him, or else justly punished according to the law Julia, which condemns those who offer violence to the public."

Anti-Machiavel:

All these ten kinds of tyrannical actions set down by Bartolus, are they not so many maxims of Machiavelli's doctrine taught to a prince? Did he not say that a prince ought to take away all virtuous people, lovers of their commonwealth; to maintain partialities and divisions; to impoverish his subjects, to nourish wars, and to do all these things which Bartolus said to be the works of tyrants? We need then no more doubt that the purpose of Machiavelli was to form a true tyrant, and that he has stolen from Bartolus one part of his tyrannical doctrine, which yet he has much augmented and enriched. For he adds that a prince ought to govern himself by his own counsel, and ought not to suffer any to discover unto him the truth of things; and that he ought not to care for any religion, neither observe any faith or oath, but ought to be cruel, a deceiver, a fox in craftiness, greedy, inconstant, unmerciful, and perfectly wicked, if it be possible, as we shall see hereafter.

~

Vindiciae contra tyrannos:

Let us then reject these detestable, faithless, and impious vanities of the court-marmosets, which make kings gods, and receive their sayings as oracles.

Anti-Machiavel:

And it seems unto me that this name of marmoset is very proper and fit for such people, and that it merits well to be called again back into use. And I believe it is drawn from hence that such people go marmoting, murmuring and whispering secretly in princes' ears flattering speeches.

~

Vindiciae contra tyrannos:

It may be the flatterers of the court will reply, that God has resigned his power unto kings, reserving heaven for himself, and allowing the earth to them to reign, and govern there according to their own fancies; briefly that the great ones of the world hold a divided empire with God himself... This discourse, I say, is worthy of the execrable Domitian who (as Suetonius recites) would be called God and Lord. But altogether unworthy of the ears of a Christian prince, and of the mouth of good subjects, that sentence of God Almighty must always remain irrevocably true, "I will not give My glory to any other," that is, no man shall have such absolute authority, but I will always remain Sovereign.

Anti-Machiavel:

The first point then, which is that the absolute power of a prince does not stretch above God, is a matter confessed by all. And there were never found any princes, or very few, who would soar and mount so high as to enterprise upon that which belonged unto God. Even the emperors Caligula and Domitian are blamed and detested by the pagan histories, which had no true knowledge of God, for that they dared enterprise upon God and that which pertained to him.

~

Vindiciae contra tyrannos:

Seeing then that kings are only the lieutenants of God, established in the Throne of God by the Lord God himself, and the people are the people of God, and that the honour which is done to these lieutenants proceeds from the reverence which is borne to those that sent them to this service, it follows of necessity that kings must be obeyed for God's cause, and not against God, and then, when they serve and obey God, and not other ways.

Anti-Machiavel:

We also see by the law of God the same absolute power is given unto kings and sovereign princes, for it is written that they shall have full power over the goods and persons of their subjects. And although God has given them their absolute power, as to his ministers and lieutenants on earth, yet he would not have the use of it but with a temperance and moderation of the second power, which is ruled by reason and equity, which we call civil.

~

Vindiciae contra tyrannos:

The Emperors Theodosius and Valentinian to Volusianus, Great Provost of the Empire.

> It is a thing well becoming the majesty of an emperor, to acknowledge himself bound to obey the laws. Our authority depending on the authority of the laws, and in very deed to submit the principality to law, is a greater thing than to bear rule. We therefore make it known unto all men, by the declaration of this our Edict, that we do not allow ourselves, or repute it lawful, to do anything contrary to this. Dated 11 June at Ravenna, under the consuls Florentius and Dionysius.

Anti-Machiavel:

This is that power which all good princes have so practiced—letting their absolute power cease without using any, unless in a demonstration of majesty, to make their estate more venerable and better obeyed—that in all their actions and in all their commands they desire to subject and submit themselves to laws and to reason... And truly all the good Roman emperors have always held this language and have so practiced their power, as we read in their histories. The emperor Theodosius made an express law for it, which is so good to be marked that I thought it good to translate it word for word.

> It is the majesty of him that governs to confess himself bound to laws, so much does our authority depend upon law. And assuredly it is a far greater thing to the empire itself to submit his empire and power unto laws. And that which we will not be lawful unto us, we show it unto others by the oracle of this our present edict. Given at Ravenna the eleventh day of June, in the year of the consulship of Florentius and Dionysius.

~

Vindiciae contra tyrannos:

Now, if they were true friends indeed, they would desire and endeavor that the king might become more powerful, and more assured in his estate according to that notable saying of Theopompus, king of Sparta, after the ephores or controllers of the kingdom were instituted. "The more" (said he) "are appointed by the people to watch over, and look to the affairs of the kingdom, the more those who govern shall have credit, and the more safe and happy the state."

Anti-Machiavel:

We read that the emperor Alexander Severus was very modest, soft, clement, and affable towards his subjects, wherewith Mammaea his mother was not content; so that one day she said unto him that he had made his authority disregarded and contemptible by his clemency. He answered, "Yea, but I have made my estate so much the longer and more assured." ...The same notable speech of Alexander is attributed to Theopompus, king of Sparta, who knew that the puissance of a king is good and excellent when kings use it well; but because there were far more kings who abused their powers, he provided for himself and his successors certain censors and correctors, which were called Ephori. Some said to Theopompus that by this establishment of Ephori he had lessened and enfeebled his power; "Nay then," he said, "I have fortified it and made it perdurable." Meaning to say, as true it is, that there is nothing which better fortifies nor which makes more firm and stable a prince's estate, than when he governs himself with such a sweet moderation that he even submits himself to the observation of laws and censures.

~

Vindiciae contra tyrannos:

But we see in many places, that when the people has despised the law, or made covenants with Baal, God has delivered them into the hands of Eglon, Jabin, and other kings of the Canaanites. And as it is one and the same covenant, so those who do break it, receive like punishment... Thou hast neglected the Lord thy God, He also has rejected thee, that thou reign no more over Israel. This has been so certainly observed by the Lord, that the very children of Saul were deprived of their paternal inheritance, for that he, having committed high treason, did thereby incur the punishment of tyrants, which affect a kingdom that in no way pertains to them. And not only the kings, but also their children and successors, have been deprived of the kingdom by reason of such felony. Solomon revolted from God to worship idols. Incontinently the prophet Ahijah foretells that the kingdom shall be divided under his son Rehoboam.

Anti-Machiavel:

David was marvelously happy in war, and always victorious over his enemies, because he was a good prince, fearing God and honoring his holy religion. Solomon his son, as long as he served God sincerely without feigning and hypocrisy, prospered very well and marvelously in a great and happy peace, and none dared stir him. But as soon as he began to practice the doctrine which Machiavelli teaches, namely to have a feigned and dissembled religion and devotion, straight had he enemies on his head, who rose up against him; as Hadad the Edomite, and Razin, who made war upon him. So generally may be said of all the kings of Judah and Israel, one after another; that God has always prospered those who were pure and sincere in religion, and who have had his service in recommendation; and contrary, upon those impure and hypocrites in religion he has heaped ruins, calamities, and other vengeances.

~

Vindiciae contra tyrannos:

Ahab, king of Israel, could not compel Naboth to sell him his vineyard; but rather if he had been willing, the law of God would not permit it.

Anti-Machiavel:

For God would not that princes use their absolute power so far as to constrain their subjects to sell their goods, as is declared to us in the example of Naboth… And hereunto agrees the divine right, whereby it is showed to us that king Ahab ought not to take away the vineyard from Naboth his subject.

~

Vindiciae contra tyrannos:

The queen Athalia, after the death of her son Ahaziah king of Judah, put to death all those of the royal blood, except little Joas, who, being yet in the cradle, was preserved by the piety and wisdom of his aunt Jehoshabeah. Athalia possesses herself of the government, and reigned six years over Judah… Finally, Jehoiada, the high priest, the husband of Jehoshabeah, having secretly made a league and combination with the chief men of the kingdom, did anoint and crown king his nephew Joas, being but seven years old. And he did not content himself to drive the Queen Mother from the royal throne, but he also put her to death, and presently overthrew the idolatry of Baal. This deed of Jehoiada is approved, and by good reason, for he took on him the defence of a good cause, for he assailed the tyranny, and not the kingdom.

Anti-Machiavel:

King Ahaziah of Judah was the son of a foreign woman named Athalia, daughter of the king of Samaria. This king governed himself by Samaritans, who were much hated by the people of Judah. At the persuasion of his mother, he gave them the principal charges and offices of his kingdom, despising and casting aside the wisest and most virtuous of his kingdom, by whom he should have governed, after the example of his predecessors. This was the cause of that king's destruction; for as Jehu was destroying the house of Ahab, he also slew Ahaziah, and exterminated almost all his race, as a partner and friend who maintained Ahab. If Ahaziah had governed himself by people of his own kingdom rather than by strangers, that evil hap would not have come to him.

~

Vindiciae contra tyrannos:

For the wisdom of a senate, the integrity of a judge, the valour of a captain, may peradventure enable a weak prince to govern well.

Anti-Machiavel:

Contrarily, if the prince be not wise at all—for it is not incompatible nor inconvenient to be a prince and to be unwise withal—yet having this resolution to govern himself by counsel, his affairs will carry themselves better than being governed by the head.

~

Vindiciae contra tyrannos:

There is, therefore, both truly mildness in putting to death some, and as certainly cruelty in pardoning of others.

Anti-Machiavel:

For a prince ought well to consider when, how, to whom, and why he pardons a fault, because it is not clemency but cruelty when a prince may do justice and does it not, as Saint Louis said.

~

Vindiciae contra tyrannos:

If the prince has committed some crime, as adultery, parricide, or some other wickedness, behold amongst the heathen, the learned lawyer Papinian who will reprove Caracalla to his face, and had rather die than obey, when his cruel prince commands him to lie and palliate his offence; nay, although he threaten him with a terrible death, yet would he not bear false witness.

Anti-Machiavel:

For briefly, a prince may well give laws unto his subjects, but it must not be contrary to nature and natural reason. This was the cause why the great lawyer Papinian, who understood both natural and civil law, loved better to die than to obey Caracalla, who had commanded him to excuse before the Senate his parricide, committed in the person of Geta his brother. For Papinian, knowing that such a crime was against natural right, would not have obeyed the emperor if he had commanded him to perpetrate it, nor would obey him so far therein as to excuse it.

Bassianus [Caracalla], not ignorant that the Senate would find this murder very strange, desired that great lawyer Papinian, his kinsman and Chancellor under Severus, to go to the Senate and make his excuses by an oration well set out: That he had done well to slay his brother, and that he had reason and occasion to do it. Papinian, who was a good man, answered that it was not so easy to excuse a parricide as it was to commit it. Bassianus, grieved at this refusal, had one of his attendants straight cut off his head.

~

Vindiciae contra tyrannos:

And instead of approving that which that villainous woman said to Caracalla, that whatsoever he desired was allowed him, we will maintain that nothing is lawful but what the law permits. And absolutely rejecting that detestable opinion of the same Caracalla, that princes give laws to others, but received none from any; we will say, that in all kingdoms well established, the king receives the laws from the people; the which he ought carefully to consider and maintain; and whatsoever, either by force or fraud he does, in prejudice of them, must always be reputed unjust.

Anti-Machiavel:

We read likewise that Caracalla, beholding one day his mother-in-law Julia with an eye of incestuous concupiscence, she said unto him, "If thou wilt, thou mayst; knowest thou not that it belongs unto thee to give the law, not to receive it?" Which talk so enflamed him yet more with lust that he took her to wife in marriage. Hereupon historiographers note that if Caracalla had known well what it was to give a law, he would have detested and prohibited such incestuous and abominable copulations, and not to have authorized them.

~

Vindiciae contra tyrannos:

Julian the apostate, did cast off Christ Jesus to cleave unto the impiety and idolatry of the pagans: but within a small time after he fell to his confusion through the force of the arm of Christ, whom in mockery he called the Galilean.

Anti-Machiavel:

The emperor Julian, who was called the Apostate, all the time of his youth, in the time of his uncle Constantine the Great, was instructed in the Christian religion; but upon a foolish curiosity he gave himself to diviners and sorcerers, to know things to come, which made him forsake Christianity… Finally after he had reigned for the space of a year and seven months, he was slain at the age of thirty-two years, making war against the Persians. Some write that as he died he blasphemed spitefully against Christ, crying "Thou hast vanquished, thou Galilean."

~

Vindiciae contra tyrannos:

Tarquinius Superbus was therefore esteemed a tyrant, because being chosen neither by the people nor the senate, he intruded himself into the kingdom only by force and usurpation.

The true causes why Tarquinius was deposed, were because he altered the custom, whereby the king was obliged to advise with the senate on all weighty affairs; that he made war and peace according to his own fancy; that he treated confederacies without demanding counsel and consent from the people or senate; that he violated the laws whereof he was made guardian; briefly that he made no reckoning to observe the contracts agreed between the former kings, and the nobility and people of Rome.

Anti-Machiavel:

Tarquin, who enterprised to slay his father-in-law king Servius Tullius to obtain the kingdom of Rome, showed well by that act and many others that he was a very tyrant... For they say that when he changed his just and royal domination into a tyrannical government, he became a contemner and despiser of all his subjects, both plebian and patrician. He brought a confusion and a corruption into justice; he took a greater number of servants into his guard than his predecessors had; he took away the authority from the Senate; moreover, he dispatched criminal and civil cases after his fancy, and not according to right; he cruelly punished those who complained of that change of estate as conspirators against him; he caused many great and notable persons to die secretly without any form of justice; he imposed tributes upon the people against the ancient form, to the impoverishment and oppression of some more than others; he had spies to discover what was said of him, and punished rigorously those who blamed either him or his government. These are the colors wherewith the histories paint Tarquin, and these are ordinarily the colors and livery of all tyrants' banners, whereby they may be known.

~

Vindiciae contra tyrannos:

Besides all this, anciently every year, and since less often, to wit, when some urgent necessity required it, the general or three estates were assembled, where all the provinces and towns of any worth, to wit, the burgesses, nobles, and ecclesiastical persons, did all of them send their deputies, and there they did publicly deliberate and conclude of that which concerned the public estate. Always the authority of this assembly was such that what was there determined, whether it were to treat peace, or make war, or create a regent in the kingdom, or impose some new tribute, it was ever held firm and inviolable; nay, which is more by the authority of this assembly, the kings convinced of loose intemperance, or of insufficiency, for so great a charge or tyranny, were disthronized

Anti-Machiavel:

Our kings of old in France used the same course that these good emperors did; for they often convocated the three Estates of the kingdom to have their advice and counsel in affairs of great consequence which touched the interest of the commonwealth. And it is seen by our histories that the general assembly of the Estates was commonly done for three causes. One, when there was a question to provide for the kingdom a governor or regent; as when kings were young, or lost the use of their understanding by some accident, or were captives or prisoners; in these cases the three Estates assembled to obtain a governor for the realm. Again, when there was cause to reform the kingdom, to correct the abuses of officers and magistrates, and to bring things unto their ancient and first institution and integrity. For kings caused the Estates to assemble, because being assembled from all parts of the kingdom, they might better be informed of all abuses and evil behaviors committed therein, and might also better work the means to remedy them; because commonly there is no better physician than he that knows well the disease and the causes thereof. The third cause why there was made an assembly of Estates was when there was a necessary cause to lay a tribute or tax upon the people; for then in a full assembly the representatives were showed the necessity of the king's and the kingdom's affairs, who graciously and courteously entreated the people to aid and help the king but with so much money as they themselves thought to be sufficient and necessary.

~

Vindiciae contra tyrannos:

About the year 1300 Pope Boniface VIII, seeking to appropriate to his See the royalties that belonged to the crown of France, Philip the Fair, the then king, did taunt him somewhat sharply: the tenor of whose tart letters are these:

> "Philip by the Grace of God, King of the French, to Boniface, calling himself Sovereign Bishop, little or no health at all. Be it known to the great foolishness and unbounded rashness, that in temporal matters we have only God for our superior, and that the vacancy of certain churches belongs to us by royal prerogative, and that appertains to us only to gather the fruits, and we will defend the possession thereof against all opposers with the edge of our swords, accounting them fools, and without brains who hold a contrary opinion."

In those times all men acknowledged the pope for God's vicar on earth, and head of the universal church. Insomuch, that (as it is said) common error went instead of a law, notwithstanding the Sorbonists being assembled, and demanded, made answer, that the king and the kingdom might safely, without blame or danger of schism, exempt themselves from his obedience, and flatly refuse that which the pope demanded; for so much as it is not the separation but the cause which makes the schism, and if there were schism, it should be only in separating from Boniface, and not from the church, nor the pope, and that there was no danger nor offence in so remaining until some honest man were chosen pope.

Anti-Machiavel:

Yet we read in our histories that our kings of France have many times hindered popes from drawing silver out of the realm, by annates, tenths, bulls, and other means; as in the time of Boniface VIII, Benedict XI, Julius II and III. But concerning this matter it is good to mark the determination made in 1410 by our masters of the faculty of the Sorbonne, and by all the University of Paris; who resolved in a general congregation that the French church was not bound to pay any silver to the pope in any manner whatsoever, unless by the way of a charitable subsidy.

As in the time of king Philip IV, Pope Boniface VIII made a decretal whereby he generally forbade all emperors, kings, and princes of Christendom to levy any tribute upon the clergy, upon pain of a present excommunication, without any other commissance or declaration. The king, because this was against his privileges (by the advice of his council, the prelates of his country, and the faculty of theology of Paris), appealed from the pope, as inferior, to the first future council, as superior.

Likewise, the pope Boniface, of whom we have spoken before, was declared a heretic by the said University and faculty of theology; not that he erred in the faith (for it was a thing whereof he had little care), but because he would needs enterprise upon the king's privileges. But as soon as he was declared a heretic, all the kingdom of France retired from his obedience.

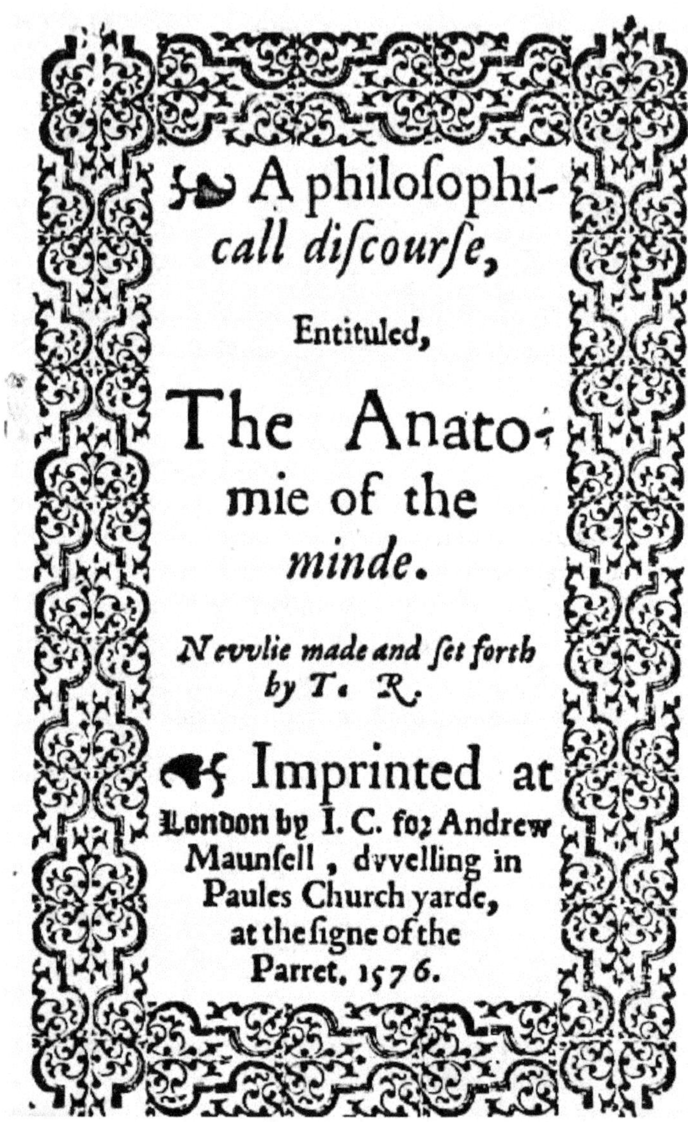

A philoſophi-
call diſcourſe,

Entituled,

The Anato-
mie of the
minde.

Nevvlie made and ſet forth
by T. R.

Imprinted at
London by I. C. for Andrew
Maunſell, dvvelling in
Paules Church yarde,
at the ſigne of the
Parret. 1576.

Appendix C: *The Anatomie of the Minde*

The Anatomie of the Minde is a small book of essays on Greek and Roman philosophy published in 1576, the year Bacon left Cambridge. It has extensive parallels with *Anti-Machiavel*, also 1576, which is somewhat strange, as the one was published in French at Geneva and the other in English at London. The book is divided into two sections, "Perturbations" and "Moral Virtues"; each chapter bears a title like Bacon's essays, "Of Ambition," "Of Constancy," etc. The front matter includes a dedication to Sir Christopher Hatton, several poems in English and Latin, and a preface. The author describes the book as "my first fruits of study" and states:

> I did once for my profit in the University draw into Latin tables, which since for thy profit (Christian Reader) at the request of a gentleman of good credit and worship, I have Englished, and published in these two books… hereafter (if God so please, and grant me life and leisure) it may be published both in sweeter phrase to delight, and in better method to profit.

Several interesting things are found here; the dedication states "Virtue though in the mind of basest for condition, is very commendable. But nothing doth so set out the Diamond, as doth the Gold . . . virtue is then most wondered at, when it is in him which for authority is of power." Bacon's essay "Of Beauty" begins "Virtue is like a rich stone, best plain set"; Shakespeare's *Richard III*:

> A base, foul stone, made precious by the foil
> Of England's chair, where he is falsely set.

The *Anatomie*'s preface states: "he which thoroughly would know himself must as well know his body, as his mind . . . For by the one we participate the nature of beasts, by the other of Angels." Compare with Bacon's essay "Of Atheism":

> They that deny a God destroy a man's nobility, for certainly man is of kin to the beasts by his body; and if he be not of kin to God by his spirit, he is a base and ignoble creature.

The *Anatomie*'s prefatory poem "Joshua Hutten to the Book" reads:

> For first the mind before old Adam's fall,
> from Perturbations all, was perfect free;
> But after, Motions and affections all,
> and passions came, which now their dwelling be.

This was how Bacon envisioned his life's work:

> Man by the Fall fell at the same time from the state of innocence and from his dominion over creation. Both of these losses, however, can even in this life be in some part repaired; the former by religion and Faith, the latter by arts and sciences.

The *Anatomie* is mentioned in Gabriel Harvey's *Pierce's Supererogation* (1593): "an anatomy of the Mind, and Fortune, were respectively as behooveful and necessary, as an Anatomy of the Body." Harvey had been Francis Bacon's rhetoric tutor at Cambridge and is credited with coining several words, including idiom, conscious, jovial, extensively, notoriety, and rascality. *Pierce's Supererogation* is also noteworthy in that it references *Venus and Adonis* before it had been published, and Harvey evidently knew it was going out under the name of Shakespeare.

Several stories found in the *Anatomie* also appear in Shakespeare; *Julius Caesar*:

> Caesar declared himself to hate and detest those which by nature were pale and sad; and therefore on a time, as he was merely jesting with many of his familiars, but especially with one of a pleasant countenance, and of constitution of body very gross, another perceiving his great familiarity, came unto him and willed him to talk not so friendly, but to take heed of him; for without doubt, he said, if he used his company and familiarity, no good would come thereof. Then Caesar smiling said that he feared not those of merry countenance, but those lowering and sad persons, meaning Brutus and Cassius; which in deed afterwards were not only the procurers, but the committers of his cruel murdering.

Julius Caesar:

> *Caesar*. Let me have men about me that are fat,
> Sleek-headed men, and such as sleep o' nights.
> Yond Cassius has a lean and hungry look.
> He thinks too much. Such men are dangerous.
> *Antony*. Fear him not, Caesar, he's not dangerous.
> He is a noble Roman, and well given.
> *Caesar*. Would he were fatter! But I fear him not.

The following scene, in which Caesar's wife Calpurnia, fearing for his life, begs him to stay home ("Your wisdom is consumed in confidence. Do not go forth today") is also anticipated in the *Anatomie*:

> Caesar likewise, being over bold and contemning the words of those which wished him well, came to a most miserable end. For oftentimes he was warned and foretold of the conspiracies of his foes to bring him to death. He was counseled to see to himself, and to guard his body, lest at any time his enemies upon the sudden should set upon him, many promising their service willingly. But he contemned all their words and would none of their service, saying that he was a miserable Prince that would have a Guard about him. But his contempt hastened his end, for as it was told him afore, his death was sought and he murdered of his Senate in their house of consultation, with penknives. If he had not so trusted to his good luck, and had such a confidence that he could have

> withstood all the assaults of his foes, and harkened to the wholesome admonitions of his faithful friends, his days might have been prolonged, and in time he might have turned the hearts of those which then were his capital and deadly enemies.

In *The Mystery of Francis Bacon* William Smedley theorized that Bacon wrote both the *Anatomie* and another (unnamed) book published shortly thereafter, which may have been *Anti-Machiavel*. Smedley did not give his reasons for essaying this attribution, but only wrote the following:

> The following suggestion is put forward with all diffidence, but after long and careful investigation. Francis Bacon was the author of two books which were published, one before he left England, and the other shortly after. The first is a philosophical discourse entitled *The Anatomie of the Minde*. "Newlie made and set forth by T.R. Imprinted at London by I.C. for Andrew Maunsell," 1576, 12mo. . . There is in existence a copy of the book with the printer's and other errors corrected in Bacon's own handwriting.

Parallelisms

Anatomie of the Minde:

The Poets feign Envy to be one of the furies of Hell, and to be fed with nothing but adders and snakes . . . The Poets feign Prometheus to be tied on the top of the mountain Caucasus, and an Eagle to be gnawing of his heart . . . the Poets feigned a notable example of Thamyras.

Bacon, *Wisdom of the Ancients*:

The poets feign that Vulcan attempted the chastity of Minerva, and impatient of refusal, had recourse to force.

Bacon, *De augmentis scientiarum*:

True history may be written in verse and feigned history in prose . . . And under the name of Poesy, I treat only of feigned history.

The division of Poesy which is aptest and most according to the propriety thereof, besides those divisions which it has in common with History (for there are feigned Chronicles, feigned Lives, and feigned Relations), is into Poesy Narrative, Dramatic, and Parabolical.

As You Like It:

No, truly; for the truest poetry is the most feigning,
and lovers are given to poetry; and what they swear in poetry may
be said as lovers they do feign

thou swear'st to me thou art honest;
now, if thou wert a poet, I might have some hope thou didst
feign

Midsummer Night's Dream:

Thou hast by moonlight at her window sung,
With feigning voice verses of feigning love

Merchant of Venice:

Their savage eyes turn'd to a modest gaze
By the sweet power of music: therefore the poet
Did feign that Orpheus drew trees, stones and floods;
Since nought so stockish, hard and full of rage,
But music for the time doth change his nature.

French Academy II:

So that we have in that part as it were a spiritual eye, which is much more excellent and profitable, then if we had bodily eyes there, as we have before, or else a face before and another behind, as the Poets feigned that Janus had.

Anti-Machiavel:

After Solon had seen Thespis' first edition and action of a tragedy, and meeting with him before the play, he asked if he was not ashamed to publish such feigned fables under so noble, yet a counterfeit personage.

Don Quixote:

Your feigned histories are so much the more good and delightful, by how much they come near the truth, or the likeness of it: and the true ones are so much the better, by how much the truer.

~

Anatomie of the Minde:

He which thoroughly would know himself must as well know his body, as his mind . . . For by the one we participate the nature of beasts, by the other of Angels.

Bacon, "Of Atheism":

They that deny a God destroy a man's nobility, for certainly man is of kin to the beasts by his body; and if he be not of kin to God by his spirit, he is a base and ignoble creature.

~

Anatomie of the Minde:

And the Pythagoreans were of his opinion, for their poesy was, that the heart should not be eaten. Their meaning was that cares and sadness should not consume the heart by unquieting the mind.

Bacon, "Of Friendship":

The parable of Pythagoras is dark, but true; *Cor ne edito*; Eat not the heart.

~

Anatomie of the Minde:

there is none either of nature so wild, or for behavior so wicked, but in their kind (as it is for a hound natural to smell, and for a bird to fly) are desirous to learn, and be cunning in somewhat.

Anti-Machiavel:

I would gladly ask this question of him that is most ignorant, vicious and carnal, whether he will not grant virtue to be a good of the soul. There is none so impudent whose conscience would not compel him to confess the same.

~

Anatomie of the Minde:

Virtue though in the mind of basest for condition, is very commendable; but nothing doth so set out the diamond, as doth the gold.

Bacon, "Of Beauty":

Virtue is like a rich stone, best plain set.

~

Anatomie of the Minde:

Cato the elder was greatly delighted with such as at the least fault would blush. And so was Diogenes the Cynic; for when talking with a young man, he perceived his face to be red with blushing, said unto him; be of good cheer my son, for this color, is the color of virtue itself.

Bacon, *The Advancement of Learning*:

It was truly said, *rubor est virtutis color* [a blush is virtue's color]

1 Henry VI:

> And, which became him like a prince indeed,
> He made a blushing cital of himself,
> And chid his truant youth with such a grace
> As if he mastered there a double spirit
> Of teaching and of learning instantly

~

Anatomie of the Minde:

Cicero says that Constancy is the health of the mind, so that by the same he understands the whole force and efficacy of wisdom, and that appears very well by her contrary. For Foolishness is nothing but a lightness and inconstancy of mind. Wherefore this constant man cannot be too much praised, seeing that either whole wisdom, or the very force of wisdom is in nothing more apparent than in Constancy.

Anti-Machiavel:

I will then presuppose that constancy is a quality which ordinarily accompanies all other virtues; it is, as it were, of their substance and nature.

Bacon, *De augmentis scientiarum*:

Constancy is the foundation on which virtues rest.

Measure for Measure:

It is virtuous to be constant in any undertaking.

Two Gentlemen of Verona:

O Heaven, were man but constant, he were perfect.

~

Anatomie of the Minde:

Homer, when he lived was of none account, every man contemned him, and none would vouchsafe to account him their countryman; but Homer being dead, was both lacked and longed for.

Coriolanus:

I shall be loved when I am lack'd.

~

Anatomie of the Minde:

Caesar declared himself to hate and detest those which by nature were pale and sad; and therefore on a time, as he was merely jesting with many of his familiars, but especially with one of a pleasant countenance, and of constitution of body very gross, another perceiving his great familiarity, came unto him, and willed him to talk not so friendly, but to take heed of him; for without doubt, he said, if he used his company and familiarity, no good would come thereof. Then Caesar smiling, said that he feared not those of merry countenance, but those lowering and sad persons, meaning Brutus and Cassius; which in deed afterwards were not only the procurers, but the committers of his cruel murdering.

Julius Caesar:

Caesar. Let me have men about me that are fat,
Sleek-headed men, and such as sleep o' nights.
Yond Cassius has a lean and hungry look.
He thinks too much. Such men are dangerous.

Antony. Fear him not, Caesar, he's not dangerous.
He is a noble Roman, and well given.

Caesar. Would he were fatter! But I fear him not.

Yet if my name were liable to fear,
I do not know the man I should avoid
So soon as that spare Cassius. He reads much,
He is a great observer, and he looks
Quite through the deeds of men. He loves no plays,
As thou dost, Antony; he hears no music.
Seldom he smiles, and smiles in such a sort
As if he mocked himself and scorned his spirit
That could be moved to smile at anything.
Such men as he be never at heart's ease
Whiles they behold a greater than themselves,
And therefore are they very dangerous.
I rather tell thee what is to be feared
Than what I fear, for always I am Caesar.

~

Anatomie of the Minde:

the Romans and many other nations allowed and thought well of [suicide]; else would not so many so desperately have bereft themselves of life; as did Brutus and Cassius after the death of Caesar; as did Antony, when he heard that Cleopatra had killed herself. For hearing the same, he broke into these words: "Die Antony, what lookst thou for? Fortune hath taken her from thee, by whom thou desired to prolong thy days, and therefore it shall never be said that such a captain as I have been accounted will be stained of a woman in stoutness of minds"; and therewithal gored himself upon a sword, and so most desperately forsook this world.

Antony and Cleopatra:

Since Cleopatra died
I have lived in such dishonor that the gods
Detest my baseness. I, that with my sword
Quartered the world, and o'er green Neptune's back
With ships made cities, condemn myself to lack
The courage of a woman – less noble mind
Than she which by her death our Caesar tells
"I am conqueror of myself."

~

Anatomie of the Minde:

Alexander the great, liberally bestowing many things upon his friends, upon a time Perdicas spoke unto him on this manner; If you thus largely still bestow your goods, O bountiful Prince, I marvel at the length, what you will keep for yourself? Then answered Alexander, for myself I reserve Hope.

Anti-Machiavel:

When Alexander the Great departed from Macedonia to go to the conquest of Asia, he had all the captains of his army appear before him, and distributed to them almost all the revenue of his kingdom, leaving himself almost nothing. One of the captains, named Perdicas, said to him: "What then will you keep for yourself?" "Even hope," answered Alexander.

Bacon, Advancement of Learning:

Lastly, weigh that quick and acute reply which [Alexander] made when he gave so large gifts to his friends and servants, and was asked what he did reserve for himself, and he answered, "Hope."

~

Anatomie of the Minde:

there is none either of nature so wild, or for behavior so wicked, but in their kind (as it is for a hound natural to smell, and for a bird to fly) are desirous to learn, and be cunning in somewhat.

Anti-Machiavel:

I would gladly ask this question of him that is most ignorant, vicious and carnal, whether he will not grant virtue to be a good of the soul. There is none so impudent whose conscience would not compel him to confess the same.

~

Anatomie of the Minde:

And therefore true is that saying of a learned man, It is hard in prosperity to know whether our friends do love us for our own sakes, or for our goods; but adversity proves a friend. For neither doth prosperity manifest a friend, nor adversity bide a flatterer.

Anti-Machiavel:

The true friend perseveres in the service of his prince, as well in time of adversity as prosperity; and the flatterer turns his back in time of adversity . . . Adversity also is a true touchstone to prove who are feigned or true friends; for when a man feels labyrinths of troubles fall on him, dissembling friends depart from him, and those who are good abide with him, as said the poet Euripides:

> Adversity the best and certain'st friends doth get,
> Prosperity both good and evil alike doth fit.

~

Anti-Machiavel:

Amity, said Cicero, is the true bond of all human society; and whoever will take amity away from among men, as Machiavelli does from among princes, he seeks to take away all pleasure, solace, contentment, and assurance that can be among humans.

Anatomie of the Minde:

Which made Cicero to say that he which would cut off this common friendship did even as it were go about to take the Sun from the world.

~

Anatomie of the Minde:

The love of our Country and Prince should be great. For (as Plato and Cicero do say) no man is born for himself, but a part of our birth our Country, a part our Parents, a part our friends challenge as due unto them.

Anti-Machiavel:

For proof hereof I will take the maxim of Plato, that we are not only born for ourselves, but that our birth is partly for our country, partly for our parents, and partly for our friends.

~

Anti-Machiavel:

as the poet Sophocles says:

> Men must not seek, nor love, of all things to get gain,

> For he that draws gain out of that which is naught,
> Before he profit gets, shall sooner loss sustain:
> For evil gotten goods are often dearly bought.

Anatomie of the Minde:

Therefore we will here conclude and say with Solon, that riches ought to be gotten, but yet after honest means, not covetously, that is by wicked arts. *Male parta, male dilabuntur*, Ill gotten goods are ill spent, says Tully.

~

Anti-Machiavel:

Finally, what mischiefs have there ever been in the world which that hideous monster perfidy has not engendered? Assuredly it is an Alecto, an infernal fury, excited and called lately from hell to the vexation and utter overthrow of this poor world.

Anatomie of the Minde:

And Sallust very notably says that by discord the greatest things come to naught; which agrees to that fiction of the Poets, who say that by discord, which is called Alecto, one of the furies of hell, the world, and all things else shall perish.

~

Anatomie of the Minde:

Scipio . . . was commonly wont to say, he had rather save one citizen than slay a thousand enemies.

Anti-Machiavel:

Here I may not forget a notable sentence of the emperor Antonius Pius, which he received from Scipio the African, which was this: That he loved better to preserve one of his subjects than to slay a thousand of his enemies.

~

Anatomie of the Minde:

It happened after that the Carthaginians being sorely foiled in battle were enforced to send Legates to Rome to entreat for peace. Hamilcar was chosen Ambassador, but calling unto mind their ill intreating of Cornelius Asina, refused to go. Then they chose Hanno, which went boldly to Rome to the Senate house, where one of the Tribunes began openly to accuse him of unfaithfulness; but the Consuls hearing thereof, commanded him to hold his peace, and said unto Hanno; Fear not, for the faithfulness of the Romans, doth rid thee from all fear of revenge; and though we have thee now in our claws, and may do with thee what we list, yet shall it not be said that treacherously we will deal with any.

Bacon, *Apophthegms*:

Hanno the Carthaginian was sent commissioner by the state, after the second Carthaginian war, to Rome, to supplicate for peace, and in the end obtained it. Yet one of the sharper senators said: You have often broken with us the peaces whereunto you have been sworn; I pray, by what Gods will you swear? Hanno answered: By the same Gods that have punished the former perjury so severely.

~

Anatomie of the Minde:

Scipio was no Philosopher by profession, but a warrior (a strange thing, that one of that sort should be so pure from unchaste cogitations), and yet being of the age of three and twenty years, and having brought under the subjection of the Romans a city in Spain, a certain Damsel without comparison among all the captives, most beautiful, was brought unto him for delectation after all his troubles. But Scipio, before he would receive any recreation at her hands, demanded what she was; which, when he understood her to be espoused unto a young man called Luceius, he thought it a shame for him to use her company beyond honesty; and so with many precious gifts and jewels, sent her safely conducted to her husband, that should be. This continence of Scipio passes all the rest. For who would think that a warrior, from a woman; a lusty young man, from a beautiful maiden; a conqueror, from a captive having time, place, and permission (so that without controlment of any man, he might have used her) would contain himself, all things falling so in the nick? And yet this noble warrior, lusty youth, and victorious conqueror, entered not familiarity with this woman, this beautiful maiden, and captive, because she had given her truth to another. O unspeakable virtue, and most wonderful continence of this noble Scipio, which so preferred honesty before lechery; chastity before incontinence; and a faithful promise, before sinful pleasure. I may not in Rhetorical manner enlarge this matter (and yet too much cannot be spoken to his praise) and therefore I leave it.

Anti-Machiavel:

Yet the example of clemency in Scipio Africanus is more notable than this of his father and uncle. After the deaths of his said father and uncle, this young lord full of all generosity and hardiness came to besiege New Carthage in Spain, and got it by assault... Among other hostages, there was a young lady of a great house brought to Scipio, who was of so great beauty that as she passed by she drew each man's regard upon her. This lady was affianced to one Allucius, prince of the Celts. Scipio, taking knowledge of her parents and to whom she was affianced, and that Allucius extremely loved her, sent for them all... The said lady's parents stepped forward and presented to him a great quantity of gold and silver for their daughter's ransom, which though Scipio refused it, they pressed it so sore upon him that he accorded to take it, and bade them lay it before him. Scipio called Allucius and said to him, Good friend, besides the dowry which your father-in-law will give you, my desire is that you will take this silver at my hands as an increase of her dowry.

French Academy:

Scipio Africanus, general of the Romans, at the taking of the city of Carthage had a young damsel taken prisoner, of rare and excellent beauty. And when he understood of what great calling she came, and how her parents not long before had betrothed her to a great lord of Spain, he commanded that he should be sent for, and restored her unto him without abusing her in any respect, although he was in the flower of his age and had free and sovereign authority. Moreover, he gave for a dowry with her the money that was brought unto him for her ransom.

~

Anatomie of the Minde:

Fabius Verruscosus (which for his virtues was called Maximus) which by circumspection did so abate the haughty courage of that victorious Hannibal, as among his friends and companions he would say that he never knew what war meant, before he had occasion to encounter with Fabius. Afterward was by the Roman Senate sent unto Fabius Maximus, Marcellus which likewise was a terror unto Hannibal. And therefore as he acknowledged Fabius to be his master, and to teach him to guide an army, so did he confess himself to stand in fear of Marcellus. Whose wisdom and circumspection was of the Romans so well noted as one of them, Fabius was called the buckler, the other Marcellus the sword (to cut off the enemies) of the people of Rome. So that as Cepio and Flaminius, for their temerity have been odious; so Fabius and Marcellus for their circumspection have been glorious in the eyes of all men.

The Advancement of Learning:

Machiavel noteth wisely, how Fabius Maximus would have been temporizing still, according to his old bias, when the nature of war was altered and required hot pursuit." Anti-Machiavel: relates that "the Roman Senate sent against Hannibal Fabius Maximus, who was not so forward (and it may be not so hardy) as Flaminius or Sempronius were; but he was more wise and careful, as he showed himself.

Bacon, *Apophthegms New and Old*:

Fabius Maximus being resolved to draw the war in length, still waited upon Hannibal's progress, to curb him; and for that purpose, he encamped upon the high grounds. But Terentius his colleague fought with Hannibal, and was in great peril of overthrow. But then Fabius came down from the high grounds, and got the day. Whereupon Hannibal said, That he did ever think, that that same cloud that hanged upon the hills, would at one time or other, give a tempest.

Anti-Machiavel:

On his arrival he did not set upon Hannibal, who desired no other thing, but began to coast him far off, seeking always advantageous places. And when Hannibal approached him, then would he show him a countenance fully determined to fight, yet always seeking places of advantage. But Hannibal, who was not so rash as to join with his enemy to his own disadvantage, made a show to recoil and fly, to draw him after him. Fabius followed him, but upon coasts and hills, seeking always not the shortest way, but that way which was most for his advantage. Hannibal saw him always upon some hill or coast near him, as it were a cloud over his head; so that after Hannibal had many times essayed to draw Fabius into a place fit for himself, and where he might give battle for his own good, and yet could not thereunto draw him, said: "I see well now that the Romans also have gotten a Hannibal; and I fear that this cloud, which approaching us, still hovers upon those hills, will one of these mornings pour out some shower on our heads."

~

Anatomie of the Minde:

And that Prince which according to reason, doth govern is called a King. So that the difference between him and a Tyrant, is because a King rules as he ought, a Tyrant as he list; a King to the profiting of all, a Tyrant only to pleasure a few, and that not for the love of virtue, but to the increase of wickedness.

Anti-Machiavel:

As contrary, none can love tyranny but must be an enemy to the common weal. For tyranny draws all to itself and despoils subjects of their goods and commodities, to appropriate all to itself, making its particular good from what belongs to all men and applying to its own profit and use what should serve for all men in general. So it follows that whoever loves the profit of a tyrant consequently hates the profit of his subjects, and he who loves the common good of subjects hates the particular profit of a tyrant.

~

Anatomie of the Minde:

Julius Caesar, though much reprehended in respect of the civil discord between him and Pompey, yet is he greatly adorned with commendations, for severe punishing the most cruel murderers of his capital enemy Pompey.

Anti-Machiavel:

In like manner was the subtle disputation of those who caused the famous captain Pompey to die. After he lost the battle of Pharsalia against Caesar, he embarked on the sea with his wife and friends, hovering about Egypt, hoping to be entertained by the young king Ptolemy in consideration of the pleasures he had done to his father. At his approach he sent a messenger to know if Ptolemy would receive him in assurance; but the king's affairs were then managed by three base persons who understood nothing less than how to govern affairs of state. They were Theodotus the rhetorician, his schoolmaster; Achillas, his domestic servant, and a chamberlain. These three venerable persons fell to counsel, to deliberate what answer the king should make to Pompey. At the beginning they differed in opinion, one saying it was good to receive him, the other not. But in the end all three accorded in the worst opinion they could have taken, which was to receive Pompey and slay him; which opinion this goodly rhetorician Theodotus persuaded to the other two by his subtle reasons. He said, "If we receive Pompey, it is certain we will have Caesar for an enemy and Pompey for a master. If we do not receive him, they will both be our enemies, Pompey for rejecting him and Caesar because we have not stayed him. But if we receive him and put him to death, Caesar will thank us and Pompey cannot revenge himself upon us; for a dead man is no warrior." Upon these goodly reasons of that subtle rhetorician, the conclusion was taken by these three bad people to put to death this great person Pompey, who had had so many triumphs and victories in his life, and who had sometimes seen five or six great kings wait on him at once, as an arbiter of their contentions and differences. If these bad counselors had considered the greatness of Pompey, who had so many virtuous and great lords as parents and friends, as also the magnanimity of Caesar, who would vanquish by true force and not by perfidies and treasons, they would never have stayed upon the cold and foolish subtleties of this gentle rhetorician, and they would not have concluded the death of so great a man. But yet they concluded it and executed their conclusion, putting Pompey to death as soon as he had taken port in Egypt. But it was not long before they received the reward of their perfidy; for Caesar soon arrived, unto whom Pothinus and Achillas presented the head of Pompey, thinking to please him greatly. Caesar turned his face away and began to weep, and commanded Pothinus and Achillas put to death. And that subtle reason of Theodotus, who persuaded them that Caesar would thank them for their murder, was not found true. Seeing this execution and finding himself very culpable, Theodotus fled and lived some years miserably wandering and begging here and there, fearing to be known by the world which everywhere had him in execration. But in the end, after the death of Caesar, Brutus found him by chance and caused him to die miserably, after he had made him endure infinite torments. Behold the end of those three counsellors of that young king Ptolemy, who also by their evil conducting made a poor end; for he was slain in a battle near the Nile, and none could ever find his body.

~

Anatomie of the Minde:

the cause of Galba the Emperor's destruction was because he lacked this Magnanimity, and suffered himself to be governed according to the minds of three wicked men, in whose company he did much delight, which brought shame to him and confusion to his people.

Anti-Machiavel:

The emperor Galba was a good and wise prince, but he suffered himself to be so governed and mastered by Titus Junius, Cornelius Lacus, and Icellus Martianus, who were of accord to rob and do evil, and brought upon Galba a common report to be a wicked and unworthy emperor. For his dealings and dispositions were not of one same tenor and constancy as they ought to have been; sometimes he showed himself too sparing, sometimes too prodigal; now remiss and negligent, now too near a taker; often he would refuse things which were not to be refused, or grant that which ought not to have been granted. He condemned noble persons upon simple suspicions; yet he would never accord to the Roman people to punish Tigellinus and Halotus, the ministers of Nero culpable of great wickedness, but contrarily favored them, and advanced Halotus into a high estate. He suffered these three counsellors and governors to sell and give tributes, freedoms, pardons for faults, and all other things. By such means Galba got the evil will of all estates, noblemen, senators, magistrates, and common people; insomuch that he was slain after reigning but seven months. And he received this end because he let himself be mastered by three alone; whereas if he had had a good council, composed of a good number of good and wise people, he would never have fallen into that misfortune; for he himself was good and wise.

~

Henry V:

Fortune is painted blind, with a muffler afore her eyes, to signify to you that Fortune is blind; and she is painted also with a wheel, to signify to you, which is the moral of it, that she is turning, and inconstant, and mutability, and variation; and her foot, look you, is fixed upon a spherical stone, which rolls, and rolls, and rolls. In good truth, the poet makes a most excellent description of it: Fortune is an excellent moral.

Bacon, "Of Fortune":

If a man look sharply and attentively, he shall see Fortune; for though she is blind, she is not invisible.

Anatomie of the Minde:

The Philosophers, and other unfaithful heathens, considering the mutability of all things, and the small assurance that man hath of anything, have supposed this world to be governed by some blind or beastly God. And hereof came the fiction of Fortune, which is of ancients, both Poets and painters feigned to be blind, brutish and frantic, and so to stand upon a round stone, distributing worldly things. She is thought to be blind, because she bestows her gifts without consideration of Persons; Brutish, because she rewards most commonly, the most ungodly; without judgement, Mad, because she is wayward, cruel and inconstant; standing not upon a square stone, for that abides, but upon a round one, for that slides continually. And therefore she is counted as brittle as glass, and nothing or more unstable. And yet notwithstanding, at her pleasure she bestows all things; which Virgil confirms, for he ascribes unto her this title Omnipotent; and Sallust says, that in all things Fortune bears sway. But let them as Heathens, and without the knowledge of the true God, imagine what they list, yet let us think, and believe none to be Omnipotent, and to dispose the world, and that which is in the same, but only our God, not Fortune; and that he does all things, not rashly without reason, but providently to our preservation; and that he is not mad in his doings, but mighty and marvelous, and doth all things to the comfort of his elect.

Anti-Machiavel:

the pagan poets have written that Fortune is a goddess who gives good and evil things to whom she will. And to denote that she does this inconsiderately and without judgment, they wrap her head in a cloth, lest with her eyes she sees and knows to whom she gives; so that she never knows unto whom she does good or evil. Moreover, they describe her standing upright upon a bowl, to denote her inconstancy, turning and tossing from side to side. Now Machiavelli would make men believe that this is true, and that all the good and evil which comes to men happens because they have Fortune accordant or discordant to their complexions. He says that she commonly favors young people, such as are hazardous and inconsiderate; to the end that thereby men might learn to be rash, violent, and heady, that they may have Fortune favorable to them. But all this doctrine tends to the same end as the former maxims do, to insinuate into men's minds and hearts a spite and utter contempt of God and his providence. For let man have once this persuasion, that no good comes to us from God, but from Fortune, he will easily forsake the service of God.

By this description of Machiavelli is evidently seen that he thinks what the poets wrote for fables concerning fortune is the very truth. For the pagan poets have written that fortune is a goddess who gives good and evil things to whom she will. And to denote that she does this inconsiderately and without judgment, they wrap her head in a cloth, lest with her eyes she sees and knows to whom she gives; so that she never knows unto whom she does good or evil. Moreover, they describe her standing upright upon a bowl, to denote her inconstancy, turning and tossing from side to side. . . For certain it is that the haps which men call Fortune proceed from God, who rather blesses prudence, which he has recommended unto us, than temerity. And although it sometimes happens that he blesses not our counsels and wisdoms, it is because we take them not from the true spring and fountain, namely from him of whom we ought to have asked it; and that most commonly we would rather our own wisdom be a glory unto us, whereas only God should be glorified.

Appendix D: *The French Academy*

L'Academie Française was published in four volumes from 1578-98 and in English translation from 1584-1618. It has numerous parallels with *The Anatomie of the Minde* and *Anti-Machiavel,* and resembles Francis Bacon's *Essays,* bearing titles "Of Ambition," "Of Hope," "Of Prosperity and Adversity," etc. As with Shakespeare's *Love's Labour's Lost,* it features four young French gentlemen secluded for purposes of study. In the dedication to Henri III, the author speaks of having attended the Estates General in 1576-7, as did Bacon. He begins: "Sir, if we credit the saying of Plato, commonwealths begin then to be happy, when kings exercise philosophy, and philosophers reign." *Anti-Machiavel*:

> There cannot come a better and more profitable thing to a people than to have a prince wise of himself; therefore, said Plato, men may call it a happy commonwealth when either the prince can play the philosopher, or when a philosopher comes to reign there.

Bacon's *Advancement of Learning*:

> Although he might be thought partial to his own profession, that said 'then should people and estates be happy, when either kings were philosophers, or philosophers kings'; yet so much is verified by experience, that under learned princes and governors there have ever been the best times.

As an example, *The French Academy* cites "Francis I, a prince of most famous memory, [who] so loved and favored letters and the professors of them that he deserved the name of the restorer of sciences and good arts" (in the Introduction we noted that the *History of the Royal Society* (1667) depicted Bacon as *Artium Instaurator,* "restorer of the arts"). *Anti-Machiavel* said "the restoration of good letters, which Francis I brought into France, did more to celebrate and immortalize his name in the memory of all Christian nations, than all the great wars and victories his predecessors had."

As with *Anti-Machiavel, The French Academy* attributes France's troubles to foreign influence:

> The ruin and destruction of this French monarchy proceeds of no other second cause (our iniquity being the first) than of the mixture which we have made of strangers with ourselves. Wherein we are not contented to seek them out under their roofs, unless we also draw them unto us and lodge them under our roofs,

> yea prefer them before our own countrymen and citizens in the offices and honorable places of this kingdom.

An English intelligence paper sometimes credited to Francis (or Anthony) Bacon, "Notes on the Present State of Christendom" (1582), reported

> division in [France] for matters of religion and state, through miscontentment of the nobility to see strangers advanced to the greatest charges of the realm, the offices of justice sold, the treasury wasted, the people polled, the country destroyed, hath bred great trouble, and like to see more.

Anti-Machiavel complains of Italian influence in France, lamenting "all France fashioned after the manners, conditions, and vices of foreigners that govern it, and who have the principal charges and estates." Shakespeare's *Richard II* laments

> Reports of fashions in proud Italy
> Whose manners still our tardy-apish nation
> Limps after in base imitation.
> Where the world doth thrust forth a vanity—
> So be it new, there's no respect how vile.

The French Academy warns: "It is a hard matter (said Socrates) for a man to bridle his desire, but he that addeth riches thereunto, is mad." *Anti-Machiavel* asked: "Who could then bridle vices and iniquities, which are fed with much wealth, and no less liberty?" Bacon's *New Atlantis* again echoes: "the reverence of a man's self is, next religion, the chiefest bridle of all vices."

Parallelisms

As You Like It:

> All the world's a stage,
> And all the men and women merely players:
> They have their exits and entrances;
> And one man in his time plays many parts,
> His acts being seven ages.

French Academy:

Now, among them that have most diligently observed the secrets of man's nature, there have been two sundry opinions concerning the division of the ages of man. Some have made seven parts, adding decrepit or bedridden age after old age, and they would ground their principal reason of this division upon this, that the number of seven is a universal and absolute number.

~

French Academy:

These are those good reasons, which ought to sound often in the ears of young men, and be supplied little by little through the study of good letters, and Moral Philosophy of ancient men, until they have wholly in possession that place of manners, which is soonest moved and most easily led, and are lodged therein by knowledge and judgement, which will be as a guard to preserve and defend that age from corruption.

Troilus and Cressida:

Not much unlike young men, whom Aristotle thought unfit to hear moral philosophy. The reasons you allege do more conduce to the hot passions of distermper'd blood than to make up a free determination 'twixt right and wrong.

Advancement of Learning:

Is not the opinion of Aristotle worthy to be regarded wherein he saith that young men are no fit auditors of moral philosophy, because they are not settled from the boiling heat of their affections, nor attempered with time and experience.

~

Much Ado About Nothing:

There was never yet philosopher
That could endure the toothache patiently,
However they have writ the style of the Gods,
And made a push at chance and sufferance.

Bacon, letter to the Earl of Essex:

It is more than a philosopher morally can digest... I esteem it like the pulling out of an aching tooth, which I remember when I was a child and had a little philosophy, I was glad when it was done.

Bacon, "Of Anger":

To seek to extinguish anger utterly is but a bravery of the Stoics. We have better oracles: Be angry, but sin not: let not the sun go down on your anger.

French Academy:

Among all the ancients, the Stoic Philosophers were most zealous and precise observers of all points concerning this virtue of patience; which they grounded upon the fatal cause of necessity, requiring such exactness and perfection thereof in men, that they would have a noble heart to be no otherwise touched with adversity than with prosperity, nor with sorrowful things than with joyful. For this cause Aristotle said that virtue only was to be wished; and therefore that it was all one to be sick or sound, poor or rich; briefly, that in all other human and necessary uses of nature, there was no more evil in one kind than in another. Whereby it seems that these Philosophers delighted in painting out a picture of such patience as never was, nor shall be among men, except first they should be unclothed of all human nature, or become as blockish and senseless as a stone.

~

French Academy:

And as the heat buried in the veins of a flint seems rather dead than alive, if the sparkles be not drawn forth by the steel, so this immortal portion of celestial fire, being the fountain and first motive of all knowledge, remains without any profit or commendable action, if it be not sharpened and set on work.

Timon of Athens:

> The fire i' the flint
> Shows not till it be struck; our gentle flame
> Provokes itself and like the current flies
> Each bound it chafes.

Troilus and Cressida:

> There were wit in this head, and 'twould
> Out" – and so there is, but it lies as coldly in him as
> fire in a flint, which will not show without knocking.

Julius Caesar:

> O Cassius, you are yoked with a lamb
> That carries anger as the flint bears fire,
> Who, much enforced, shows a hasty spark,
> And straight is cold again.

~

French Academy:

Thales said that nothing in all the world was more common than Hope, because it abides with them also that have no other goods.

Measure for Measure:

> The miserable have no other medicine but only hope.

~

French Academy:

Samson betrayed by Dalilah, Solomon became brutish through his concubines, Ahab rooted out through Jezebel, Mark Antony slew himself for the love of Cleopatra, the destruction of Troy because of Helena, the Pandora of Hesiod, the pitiful death of Hercules by Deiamra, and many other miserable events procured chiefly by women, and plentifully declared in histories.

Love's Labour's Lost:

Love is a devil. There is not evil angel but Love. Yet was Samson so tempted, and he had an excellent strength; yet was Solomon so seduced, and he had a very good wit. Cupid's butt shaft is too hard for Hercules' club, and therefore too much odds for a Spaniard's rapier.

~

French Academy:

Thales, one of the Sages of Greece, minding to show that it was not good for a man to marry when one asked him why he married not, being in the flower of his age, said that it was not yet time. Afterward, being grown to further age and demanded the same question, he answered, that the time was past.

Bacon, *Apophthegms*:

Thales being asked when a man should marry, said young men not yet, old men not at all.

~

French Academy:

And to those whose minds are not well disposed, neither riches, nor strength, nor beauty can be judged good, but the greater increase arises of them, the more harm they may procure to him that possesses them

As You Like It:

Know you not, master, to some kind of men their graces serve them but as enemies? No more do yours; your virtues, gentle master, are sanctified and holy traitors to you. Oh, what a world is this, when what is comely envenoms him that bears it!

~

French Academy:

Was there ever any captain among the Romans greater than Julius Caesar? Yet was he of a weak and tender complexion, subject to great headaches, and visited sometimes with the falling sickness.

Julius Caesar:

> *Casca*. He fell down in the market-place, and foam'd at the mouth, and was speechless.
>
> *Brutus*. 'Tis very like; he hath the falling-sickness.

~

As You Like It:

> Your "if" is the only peacemaker; much virtue in "if."

French Academy:

And when they were to answer anything propounded on a sudden, avoiding all superfluous speech, their answers were very witty and well contrived, their words very significant and short, having in them both grace and gravity joined together. As when Philip king of Macedonia wrote unto them, that if he entered within Laconia, he would overthrow them topsy-turvy; they wrote back unto him only this word, *If*.

~

French Academy:

This was wisely noted by Philippides, when Lysimachus the king asked him what part of his goods he would have imparted unto him. "What you please Sir" (said he), "so it be no part of your secrets."

Pericles:

Well, I perceive he was a wise fellow and had good discretion that, being bid to ask what he would of the King, desired he might know none of his secrets.

~

Troilus and Cressida:

> They tax our policy and call it cowardice,
> Count wisdom as no member of the war,
> Forestall prescience, and esteem no act
> But that of hand. The still and mental parts
> That do contrive how many hands shall strike
> When fitness calls them on and know by measure
> Of their observant toil the enemy's weight –
> Why, this hath not a finger's dignity.
> They call this bed work, mappery, closet war;
> So that the ram that batters down the wall,
> For the great swinge and rudeness of his poise,
> They place before his hand that made the engine,

> Or those that with the fineness of their souls
> By reason guide his execution.

French Academy:

Fabius the Greatest comes first to my remembrance, to prove that the resolution of a courageous heart, grounded upon knowledge and the discourse of reason, is firm and immutable. This Captain of the Roman army, being sent into the field to resist the fury and violence of Hannibal, who being Captain of the Carthaginians, was entered into Italy with great force, determined for the public welfare and necessity to delay and prolong the war, and not to hazard a battle but with great advantage. Whereupon certain told him that his own men called him Hannibal's schoolmaster, and that he was jested at with many other opprobrious speeches, as one that had small valor and courage in him; and therefore they counseled him to fight, to the end he might not incur any more such reprehensions and obloquies. I should be (said he again to them) a greater coward than now I am thought to be, if I should forsake my deliberation necessary for the common welfare and safety, for fear of their girding speeches and bolts of mockery, and obey those (to the ruin of my country) whom I ought to command. And indeed afterward he gave great tokens of his unspeakable valor, being sent with three hundred men only to encounter with the said Hannibal; and seeing that he must of necessity fight for the safety of the Commonwealth, after all his men were slain, and himself hurt to death, he rushed against Hannibal with so great violence and force of courage, that he took from him the diadem or frontlet, which he had about his head, and died with that about him.

~

French Academy:

As the remembrance of an evil is kept a long time, because that which offends is very hardly forgotten, so we commonly see that the memory of benefits received is as suddenly vanished and lost, as the fruit of the good turn is perceived.

Troilus and Cressida:

> Time hath, my lord, a wallet at his back,
> Wherein he puts alms for oblivion,
> A great-sized monster of ingratitudes.

Those scraps are good deeds past, which are devoured
As fast as they are made, forgot as soon
As done.

~

French Academy:

This is that which Possidonius teaches us, saying that anger is nothing else but a short fury.

Timon of Athens:

They say, my lords, "*Ira furor brevis est*"; But yond man is ever angry.

~

French Academy:

Timon the Athenian, detesting much more than all these the imbecility of man's nature, used and employed all his skill to persuade his countrymen to abridge and shorten the course of their so miserable life, and to hasten their end by hanging themselves upon gibbets, which he had caused to set up in great number, in a field that he bought for the same purpose, unto whose persuasions many gave place.

Timon of Athens:

Tell my friends,
Tell Athens, in the sequence of degree
From high to low throughout, that whoso please
To stop affliction, let him take his haste,
Come hither ere my tree hath felt the ax,
And hang himself.

~

French Academy:

But to the end we confound not together that which is simply divine, with that which is human, I think we ought to make a double hope; the first true, certain, and infallible, which concerns holy and sacred mysteries; the other doubtful, respecting earthly things only.

Troilus and Cressida:

> The ample proposition that hope makes
> In all designs begun on earth below
> Fails in the promised largeness. Checks and disasters
> Grow in the veins of actions highest reared,
> As knots, by the conflux of meeting sap,
> Infects the sound pine and diverts his grain
> Tortive and errant from his course of growth.

~

French Academy:

Sir, if we credit the saying of Plato, commonwealths begin then to be happy, when kings exercise philosophy, and philosophers reign.

Anti-Machiavel:

I am content to presuppose that it is certain that there cannot come a better and more profitable thing to a people than to have a prince wise of himself; therefore, said Plato, men may call it a happy commonwealth when either the prince can play the philosopher, or when a philosopher comes to reign there.

~

French Academy:

The desire which Plato had to profit many caused him to sail from Greece into Sicily, that by grave discourses and wise instructions he might stay and contain within the bounds of reason the young years of Dionysius, prince of that country, who through unbridled liberty and power not limited waved hither and thither without restraint. Afterward, when he began to be in love with the beauty of learning, he left off little by little his drunkenness, maskings, and whoredoms, wherein before he gloried; insomuch that his court was wholly changed upon a sudden, as if it had been inspired from heaven. But within a while after, Dionysius giving ear to flatterers, banished Plato; to whom when he took his leave of him, the tyrant said, I doubt not, Plato, but you will speak ill of me when you are in the University among thy companions and friends. Whereupon the philosopher smiling and observing that freedom of speech which he had always used towards him,

made this answer. I pray God, Sir, there may never be so great want of matter to speak of in the University, that we need to speak of thee.
Anatomie of the Minde:

It is written of Dionysius, a most cruel tyrant, that as long as he perceived himself to be well reported of, he was a good man, but when the privy talk to his defamation came to his ears, he then began to leave his good nature, and to exercise all kind of cruelty toward his subjects, and became the most cruel Prince that ever was.

~

French Academy:

It is a hard matter (said Socrates) for a man to bridle his desire, but he that addeth riches therunto, is mad.

Anti- Machiavel:

Who could then bridle vices and iniquities, which are fed with much wealth, and no less liberty?

~

French Academy:

He that has but half an eye may see that there are a great many amongst us of those foolish men of whom David speaks, *Who say in their hearts that there is no God*. In the forefront of which company, the students of Machiavel's principles and practicers of his precepts may worthily be ranged. This bad fellow, whose works are no less accounted of among his followers than were Apollo's oracles amongst the heathen, nay than the sacred Scriptures are among sound Christians, blushed not to belch out these horrible blasphemies against pure religion, and so against God the author thereof…

Anti-Machiavel:

For what shall I speak of religion, whereof the Machiavellians had none, as already plainly appears; yet they greatly labored also to deprive us of the same… he is of no reputation in the court of France who has not Machiavelli's writings at the fingers' ends, both in the Italian and French tongues, and can apply his precepts to all purposes, as the oracles of Apollo.

~

French Academy:

For the ruin and destruction of this French monarchy proceeds of no other second cause (our iniquity being the first) than of the mixture which we have made of strangers with ourselves. Wherein we are not contented to seek them out under their roofs, unless we also draw them unto us and lodge them under our roofs, yea prefer them before our own countrymen and citizens in the offices and honorable places of this kingdom… they have left us nothing but new manners and fashions of living in all dissoluteness and pleasure; except this one thing also, that we have learned of them to dissemble, and withal to frame and build a treason very subtly. Such is the provision wherewith our French youth is commonly furnished by their Italian voyages.

Anti-Machiavel:

For besides the examples we read in histories, we know it by experience, seeing at this day all France fashioned after the manners, conditions, and vices of foreigners that govern it, and who have the principal charges and estates. And not only many Frenchmen are such beasts to conform themselves to strangers' complexions, but also to gaggle their language and disdain the French tongue as a thing too common and vulgar.

~

French Academy:

This is that which at length (as Crates the philosopher said very well) stirs up civil wars, seditions, and tyrannies within cities; to the end that such voluptuous men, and ambitious of vainglory, fishing in a troubled water, may have wherewith to maintain their foolish expenses, and so come to the end of their platforms.

Anti-Machiavel:

And would to God that the French nation had never been of that nature and condition to do well unto strangers, without first knowing and trying their behaviors and manner of life. We should not then see France to be governed and ruled by strangers, as it is; we should not feel the calamities and troubles of civil wars and dissentions, which they enterprise to maintain their greatness and magnitude, and to fish in troubled water.

~

French Academy:

But whatsoever my speech has been hitherto, my meaning is not to find fault with the right use of hospitality, which ought to be maintained and kept inviolable in every well-established commonwealth. In this respect France has been commended above all nations for entertaining and receiving all sorts of people; provided always that they be not preferred before our own children, and that they be contented to obey and live according to the common laws of the country.

Anti-Machiavel:

For hospitality is recommended unto us by God, and it is a very laudable virtue for men to entertain strangers and entertain them well; but strangers also ought to content themselves to be welcomed and entertained in a country or town, without aspiring to master or hold offices and estates. The French nation is that which of all Christendom receives and loves strangers most, for they are as welcome all over France as those of their own nation.

~

French Academy:

What ought they to do, that say they are all members of that one head, who recommends so expressly unto them meekness, mildness, gentleness, grace, clemency, mercy, good will, compassion, and every good affection towards their neighbor? All which things are comprehended under this only sacred word of Charity.

Anti-Machiavel:

True charity is joined unto faith, pity, and all other virtues

~

French Academy:

Notwithstanding, wisely applying themselves to places and persons, they can in their serious discourses intermingle some honest pastimes, but yet not altogether without profit. As Plato in his foresaid feast interlaces certain comical speeches of love, howbeit all the rest of the supper there was nothing but wise discourses of philosophy.

Anti-Machiavel:

But seeing we are entered into this talk, we will look deeper into the matter to draw out some good resolution from this question, by the way only of a tentative and pleasant disputation, and not of a full determination hereof. For as Cato says, amongst serious things joyous and merry things would be sometimes mixed.

~

French Academy:

Kings, princes and magistrates, who because they see and hear for the most part by other men's eyes and ears, ought necessarily to have such friends, counsellors, and servants about them, as will freely tell them the truth, as hereafter we may discourse more at large.

Anti-Machiavel:

And to attend while the prince himself begins the matter first to his council, would be in vain; for he cannot propose what he does not know, and it is a notorious and plain thing that the prince, who is always shut up in a house or within a troupe of his people, sees not nor knows how things pass, but what men make him see and know.

~

French Academy:

Francis I, a prince of most famous memory, so loved and favored letters and the professors of them that he deserved the name of the restorer of sciences and good arts, sparing neither care nor means to assemble together books and volumes of sundry sorts and of all languages for the beautifying of his so renowned a library, which was a worthy monument of such a magnifical Monarch; whose praiseworthy qualities we see revived in our king, treading in the selfsame steps.

Anti-Machiavel:

We see that the restoration of good letters, which Francis I brought into France, did more to celebrate and immortalize his name in the memory of all Christian nations, than all the great wars and victories his predecessors had... In our time Francis I imitated the example of this great and wise emperor, establishing public lectures at great wages in the University of

Paris, a thing whereof his memory has been and shall be more celebrated through the world than for so many great wars he valiantly sustained during his reign... You have gloriously crowned that work, which that great king Francis your grandfather did happily begin, to the end that arts and sciences might flourish in this kingdom.

~

French Academy:

It is a usual speech in the mouths of men altogether ignorant of the beauty and profit of Sciences, that the study of letters is a bottomless gulf, and so long and uneasy a journey that they who think to finish it, oftentimes stay in the midway, and many being come to the end thereof find their minds so confused with their profound and curious skill, that instead of tranquility of soul, which they thought to find, they have increased the trouble of their spirit.

Anti-Machiavel:

For there are at this day infinite persons who so much please themselves in profane authors, some in poets, some in historiographers, some in philosophy, some in physic, or in law, that they care nothing to read or else to know anything for the salvation and comfort of their souls. Some care not at all for it, others reserve that study until they have ended the studies of other sciences, and in the meanwhile the time runs away, and often it comes to pass that when they leave this world, their profane studies are not ended, nor the study of holy letters commenced, and so they die like beasts.

~

French Academy:

Through want of skill and ignorance he falls into a worse estate than he was in before, and as we commonly say, from a gentle ague into a pestilent and burning fever.

Anti-Machiavel:

They were fallen from a shaking fever into a hot ague, as the French proverb is.

~

French Academy:

Whereunto also the precepts and discourses of learned and ancient philosophers may serve for our instruction and pricking forward; as also the examples (which are lively reasons) of the lives of so many notable men, as histories, the mother of antiquity, do as it were represent alive before our eyes.

Anti-Machiavel:

And you, good Edward, imitate the wisdom, sanctimony, and integrity of your father, the Right Honorable Lord Nicholas Bacon, Keeper of the broad Seal of England, a man right renowned; that you may lively express the image of your father's virtues in the excellent towardness which you naturally have from your most virtuous father. If you both daily ruminate and remember the familiar and best known examples of your ancestors, you cannot have more forcible persuasions to move you to that which is good and honest.

~

French Academy:

If we compare worldly goods with virtue (calling that good which usurps that name, and is subject to corruption); first, as touching those which the philosophers call the goods of fortune, and namely nobility, whereon at this day men stay so much; what is it but a good of our ancestors?

Anti-Machiavel:

I will also note another notable vice which runs current among gentlemen at this day, which is that they make so great account of their nobility of blood that they esteem not the nobility of virtue; insomuch that it seems to some that no vices can dishonor or pollute the nobility and gentry which they have from their ancestors. But they ought well to consider that to their race there was a beginning of nobility, which was attributed to the first that was noble in consideration of some virtue that was in him.

~

French Academy:

Ambition truly is the most vehement and strongest passion of all those wherewith men's minds are troubled; and yet many notable and virtuous men have so mastered it by the force of their temperance that oftentimes they accepted offices and estates of supreme authority, as it were by compulsion and with grief; yea some altogether contemned and willingly forsook them.

Anti-Machiavel:

Besides all this, in the election of counsellors and magistrates he did ever suspect those who sought offices, and held them for ambitious and dangerous people to the common weal. But they who he could know to be good men and worthy of public charge, and never sought it, these were they who he esteemed most sufficient; and the more they excused themselves from accepting offices, so much the more were they constrained unto them.

~

French Academy:

The custom that Aurelius Severus used is much more praiseworthy. For when he sent governors into the provinces, he caused their names to be published many days before, to the end that whosoever knew anything in them worthy of reprehension, he should give notice thereof; and they that reported truly, were promoted to honor by him and slanderers grievously punished.

Anti-Machiavel:

And upon that point, it seems to me that the manner of proceeding which Alexander Severus used to choose his counsellors and his magistrates, is very good and merits well to be imitated and drawn into consequence... And the better to be informed of the reputation of persons whereof he had proffers by his wise friends, he caused to be set up in common streets and great public areas, where many ways meet, certain posts to fix bills upon them, whereupon was written certain exhortations unto the people, that if any man had anything to say against such and such a man (which he named) wherefore they might not be received and admitted to such and such an office, that he should denounce it. And so made those commands by placards, to the end he might better discover and be advertised of the virtues and vices of persons.

~

French Academy:

Caligula, a most cruel emperor, never had secure and quiet rest, but being terrified and in fear awoke often, as one that was vexed and carried headlong with wonderful passions. Nero, after he had killed his mother, confessed that while he slept he was troubled by her, and tormented with Furies that burned him with flaming torches.

Anti-Machiavel:

What repose could Nero have, who confessed that often the likeness of his mother, whom he slew, appeared to him, which tormented and afflicted him; and that the furies beat him with rods and tormented him with burning torches. What delicateness or sweetness of life could Caligula and Caracalla have? who always carried coffers full of all manners of poisons, as well to poison others as themselves in case of necessity, for fear they should fall alive into the hands of their enemies.

~

Anti-Machiavel:

The governor of Judea, called Petronius, would have placed an image of Caligula in the great temple of Jerusalem; but the Jews, who extremely detested images, would not suffer him; whereby there was likely to have been a great sedition.

French Academy:

Caligula, a Roman emperor, sent Petronius into Syria with commandment to make war with the Jews if they would not receive his image into their temple. Which when they refused to do, Petronius said unto them that then belike they would fight against Caesar, not weighing his wealth or their own weaknesses and inability.

~

French Academy:

Alexander the Great, being by the states of all Greece chosen general captain to pass into Asia and to make war with the Persians, before he took ship he inquired after the estate of all his friends to know what means they had to follow him. Then he distributed and gave to one lands, to another a village, to this man the custom of some haven, to another the profit of some borough town, bestowing in this manner the most part of his demeans and revenues. And when Perdicas, one of his lieutenants, demanded of him what he reserved for himself: he answered Hope.

Anti-Machiavel:

When Alexander the Great departed from Macedonia to go to the conquest of Asia, he had all the captains of his army appear before him and distributed to them almost all the revenue of his kingdom, leaving himself almost nothing. One of the captains, named Perdicas, said to him: "What then will you keep for yourself?" "Even hope," answered Alexander.

~

French Academy:

Anaxarchus the philosopher, being taken prisoner by the commandment of Nero, that he might know of him who were the authors of a conspiracy that was made against his estate, and being led towards him for the same cause, he bit his tongue in sunder with his teeth and spit in his face, knowing well that otherwise the tyrant would have compelled him by all sorts of tortures and torments to reveal and disclose them. Zeno missing his purpose, which was to have killed the tyrant Demylus, did as much to him.

The Anatomie of the Minde:

Zeno the Stoic being cruelly tormented of a King of Cypress to utter those things which the king was desirous to know, at length because he would not satisfy his mind, bit off his own tongue, and spit the same in the tormentor's face. But the constancy of Anaxarchus was more strange, for being taken of Nicocreon, a most cruel of all other Tyrants, and afterward hearing that by the commandment of the Tyrant he should in a mortar he bruised and broken into pieces, said most constantly unto him in this manner; Bruise and break this body of mine at thy pleasure O Tyrant, yet shalt thou never diminish any whit of Anaxarchus. Then the Tyrant because he could not abide his bold speech, commanded that his tongue should be cut out of his mouth. But Anaxarchus laughing at his madness, thought he should never have his mind, and therefore he bit out his own tongue and spit the same by mammocks upon the tyrant's face.

~

Anatomie of the Minde:

what so savage as Xerxes, which appointed a great reward unto him, which invented a new pleasure never heard of before?

French Academy:

Xerxes, monarch of the Persians, was so intemperate and given to lust that he propounded rewards for those that could invent some new kind of pleasure; and therefore coming into Greece with an infinite number of men to subdue it, he was overcome and repulsed by a small number, as being an effeminate and fainthearted man.

~

French Academy:

Fabius the Greatest comes first to my remembrance, to prove that the resolution of a courageous heart, grounded upon knowledge and the discourse of reason, is firm and immutable. This captain of the Roman army being sent into the field to resist the fury and violence of Hannibal, who being captain of the Carthaginians, was entered into Italy with great force, determined for the public welfare and necessity to delay and prolong the war, and not to hazard a battle but with great advantage.

Anti-Machiavel:

Seeing this, the Roman Senate sent against Hannibal Fabius Maximus, who was not so forward (and it may be not so hardy) as Flaminius or Sempronius were; but he was more wise and careful, as he showed himself. On his arrival he did not set upon Hannibal, who desired no other thing, but began to coast him far off, seeking always advantageous places. And when Hannibal approached him, then would he show him a countenance fully determined to fight, yet always seeking places of advantage. But Hannibal, who was not so rash as to join with his enemy to his own disadvantage, made a show to recoil and fly, to draw him after him. Fabius followed him, but upon coasts and hills, seeking always not the shortest way, but that way which was most for his advantage.

~

French Academy:

Scipio Africanus, general of the Romans, at the taking of the city of Carthage had a young damsel taken prisoner, of rare and excellent beauty. And when he understood of what great calling she came, and how her parents not long before had betrothed her to a great lord of Spain, he commanded that he should be sent for, and restored her unto him without abusing her in any respect, although he was in the flower of his age and had free and sovereign authority. Moreover, he gave for a dowry with her the money that was brought unto him for her ransom.

Anti-Machiavel:

Yet the example of clemency in Scipio Africanus is more notable than this of his father and uncle. After the deaths of his said father and uncle, this young lord full of all generosity and hardiness came to besiege New Carthage in Spain, and got it by assault... Among other hostages, there was a young lady of a great house brought to Scipio, who was of so great beauty that as she passed by she drew each man's regard upon her. This lady was affianced to one Allucius, prince of the Celts. Scipio, taking knowledge of her parents and to whom she was affianced, and that Allucius extremely loved her, sent for them all... The said lady's parents stepped forward and presented to him a great quantity of gold and silver for their daughter's ransom, which though Scipio refused it, they pressed it so sore upon him that he accorded to take it, and bade them lay it before him. Scipio called Allucius and said to him, Good friend, besides the dowry which your father-in-law will give you, my desire is that you will take this silver at my hands as an increase of her dowry.

~

French Academy:

Camillus, a Roman dictator, is no less to be commended for that which he did during the siege of the City of the Fallerians. For he that was schoolmaster of the chiefest men's children among them, being gone out of the city, under color to have his youth to walk and to exercise themselves along the walls, delivered them into the hands of the Roman captain; saying unto him that he might be well assured the citizens would yield themselves to his devotion, for the safety and liberty of that which was dearest unto them. But Camillus, knowing this to be too vile and wicked a practice, said to those that were with him, that although men used great outrage and violence in war, yet among good men certain laws and points of equity were to be observed. For victory was not so much to be desired, as that it should be gotten and kept by such cursed and damnable means; but a general ought to war, trusting to his own virtue, and not to the wickedness of others. Then stripping the said schoolmaster, and bending his hands behind him, he delivered him naked into the hands of his scholars, and gave to each of them a bundle of rods, that so they might carry him back again into the city. For which noble act the citizens yielded themselves to the Romans, saying that in preferring justice before victory, they had taught them to choose rather to submit themselves unto them, than to retain still their liberty; confessing withal that they were overcome more by their virtue than vanquished by their force and power.

Anti-Machiavel:

Camillus, a Roman general, besieged the town of Falisques, the Romans' enemies. The schoolmaster of Falisques enterprised a great wickedness and villainy; for making a countenance to lead, for sport and pastime, the youth of that town who were committed to him to be instructed, he straight brought them to Camillus' camp, hoping he would give some good recompense, speaking in this manner. "Lord Camillus, I yield into your hands the town of Falisques, for I here bring you their dear and loving children, which to recover they will easily yield themselves to you." To whom Camillus answered, "Wicked wretch, you do not address yourself to your like. We have by compacts no society with the Falisques, but by nature we have; we are not ignorant of the right of war and of peace, which we will courageously observe. We make not war upon young children, for even when we take towns, we pardon them, so do we also to them who bear arms against us. You would vanquish the Falisques by deceit and villainy, but I will vanquish them by virtue and arms, as I overcame the Veians." After this, Camillus commanded to bind the schoolmaster's hands behind him, and to give all the scholars rods in their hands, who whipped him naked into the town. As thus in this sort the children brought their master to the town, all the people ran to see the spectacle; which so changed their courage, before full of wrath and hatred against the Romans, that they straight sent delegates to Camillus to desire peace, admiring the Roman clemency and justice. Camillus, knowing that he alone could not enterprise to conclude a peace, sent the delegates towards the Senate of Rome, where on arriving they made this speech. "My masters, having been vanquished by an agreeable victory both to gods and men, we yield ourselves to you, knowing that our estate shall be better under your domination than in our own liberties and customs. The issue of this war will serve hereafter for a double example to all mankind, for it seems you better love loyalty in war than present victory. And we, being provoked by your kindness and loyalty, gladly and willingly yield you the victory. We offer ourselves your subjects, and we shall never repent ourselves of your domination, nor you of your loyalty." The peace and alliance accorded to the Falisques, Camillus entered Rome in triumph, and was more esteemed to be a victor by clemency than if it had been by arms.

~

French Academy:

Caracalla the emperor, traveling with his army towards the Parthians, under pretense of marrying the daughter of Artabanus their king, who came for the same purpose to meet him, he set upon him contrary to his faith, and put him to flight with an incredible murder of his men. But within a while after, being come down from his horse to make water, he was slain by his own men; which was noted as a just punishment sent from God for his unfaithfulness.

Anti-Machiavel:

[Caracalla] also played another part of treachery, under the pretext and show of marriage, with Artabanus, king of the Parthians. For he wrote letters to him whereby he signified that their empires were the two greatest empires of the world; and that being the son of a Roman emperor, he could not find a better wife than the daughter of the king of the Parthians. He therefore asked her hand in marriage, to join the greatest empires of the earth and to end their wars. The king at first denied Caracalla his daughter, saying that such a marriage was very unfit because of the diversity of their languages, manners, and habits; also because the Romans had never before allied or married with the Parthians. But upon this refusal Caracalla insisted and pressed him more strongly than before, and sent to Artabanus great gifts, so that in the end he gave to him his daughter. Caracalla, assuring himself that he would find no hostility in the Parthian country, boldly entered far into it with his army, saying he went but to see the king's daughter. On the other side, Artabanus prepared himself and his retinue in as good order as was possible, without any army, to go meet his new son-in-law. What did this perfidious Caracalla? As soon as the two parties met, and Artabanus came near to salute and embrace him, he commanded his soldiers to charge upon the Parthians. The Romans attacked as if there had been an assigned battle, and there was a great slaughter made of the

Parthians; but the king, with the help of a good horse, escaped with great difficulty and danger. He determined to revenge himself of that villainy and treachery; but Macrinus relieved him of that pain, and soon slew that monster Caracalla, who was already detested through all the world because of his perfidy.

~

French Academy:

Antoninus and Geta, brothers and successors in the empire to Severus their father, could not suffer one another to enjoy so large a monarchy; for Antoninus slew his brother Geta with a dagger, that himself might rule alone.

Anti-Machiavel:

Severus intending to leave the government of the empire to his two sons together, flatterers about them disposed it otherwise… Those two young princes fell into so great and mortal enmity, that not only they hated all each other's friends and servants, but also those who would have reconciled them. As soon as their father Severus was dead, Laetus, one of the marmosets of Bassianus, persuaded him to slay his brother and feign that he was assailed by him. This counsel was found good by Bassianus, who was audacious enough and ready to give the blow with his own hand. One morning he entered into the chamber of empress Julia, Geta's mother, and finding him there he slew him between his mother's arms.

~

French Academy:

And it is greatly to be feared that such unskillful and ambitious men will in the end show themselves both in will and practice to be imitators of one Cleander, an outlandish slave, who being preferred by Commodus the emperor to goodly offices and great places of honor, as to be great master of his men of war and his chief chamberlain, conspired notwithstanding against his lord, seeking to attain to the imperial dignity by seditions which he stirred up in Rome between the people and the soldiers. But through good order taken, his enterprise took no effect, except the loss of his own head and destruction of his house.

Anti-Machiavel:

Cleander was another marmoset who succeeded in his place; who at the beginning made some show that he would do better, but soon did worse. He practiced many cruelties, and sold the estates and governments of provinces to those who would offer most. There happened at Rome then a great famine and pestilence. The people, who always lay the cause of public calamity upon the governors, bruited abroad that Cleander was the cause of the plague and the famine, and therefore should die. Cleander, to stop this rumor and cause the people to hold their peace, had the emperor's horsemen rush through the town and suburbs, slaying and wounding innumerable. But the people began to take houses and fight from the windows so well that the horsemen were constrained to retire. Fadilla, the sister of Commodus, seeing this civil war commenced and raised by Cleander, went to find her brother, whom she found among his harlots. All bewept she fell on her knees before him, saying, "Sir, my brother, you are here taking your pleasures, and know not the things that pass, nor the danger wherein you are. For both yours and our blood is in peril, to be altogether exterminated by the war and civil stir Cleander has raised in the town. He has armed your forces, and has made them rush against the people, and has brought them unto a slaughter more than barbarous, filling the streets with Roman blood. If you do not soon put to death the author of this evil, the people will fall upon you and us, and tear us in pieces." Saying these words, she tore her garments and was very sad, as it were desperate. Many also who were present increased the fear of Commodus by their persuasions; fearing some great danger to himself, he sent in haste for Cleander, who knew nothing of his complaint. As soon as he arrived, Commodus had his head cut off and carried on a pike through the town, and the sight of the head appeased the people.

~

French Academy:

And lastly, for the upshot and perfection of all happiness and felicity in this world, he instructs him how he may lead a quiet and peaceable life in beholding the wonderful works of the divinity, which he is to adore and honor, and in the amendment and correction of his manners naturally corrupted, by squaring them after the pattern of virtue, that so he may be made worthy and fit to govern human affairs, for the profit of many; and at length attain to the perfection of a wise man, by joining together the active life with the contemplative in the certain hope and expectation of a second, immortal and most blessed life.

Anti-Machiavel:

Very true it is, that among Christians there must be some contemplatives, that is to say, studious people who give themselves to holy letters in order to teach others. But we do not find by the documents of that religion that there is allowed any idle contemplation of dreamers, who do nothing but imagine dreams and toys in their brains; but a contemplative life of laboring studious people is only approved, who give themselves to letters to teach others. For after they have accomplished their studies, they ought to put in use and action that which they know, bringing into an active life that which they have learned by their study in their contemplative life. And those who use this otherwise do not follow the precepts of the true Christian religion.

~

French Academy:

For this cause the ancient Romans built two temples joined together, the one being dedicated to virtue and the other to honor; but yet in such sort that no man could enter into that of honor except first he passed through the other of virtue.

Anatomie of the Minde:

Marcellus building a temple, which he called the temple of Honor, did so place and situate the same, as none could have entrance thereunto, except first he came through Virtue's temple. Signifying thereby, that the way to honor, is by virtue only, not by favor, money, nor other means.

www.ingramcontent.com/pod-product-compliance
Lightning Source LLC
LaVergne TN
LVHW050620100826
845148LV00011B/1666

* 9 7 9 8 3 8 5 2 7 8 7 7 0 *